Moons: A Very Short Introduction

VERY SHORT INTRODUCTIONS are for anyone wanting a stimulating and accessible way into a new subject. They are written by experts, and have been translated into more than 40 different languages.

The series began in 1995, and now covers a wide variety of topics in every discipline. The VSI library now contains over 400 volumes—a Very Short Introduction to everything from Psychology and Philosophy of Science to American History and Relativity—and continues to grow in every subject area.

Very Short Introductions available now:

ACCOUNTING Christopher Nobes
ADVERTISING Winston Fletcher
AFRICAN AMERICAN RELIGION
    Eddie S. Glaude Jr.
AFRICAN HISTORY John Parker and
    Richard Rathbone
AFRICAN RELIGIONS
    Jacob K. Olupona
AGNOSTICISM Robin Le Poidevin
ALEXANDER THE GREAT
    Hugh Bowden
ALGEBRA Peter M. Higgins
AMERICAN HISTORY Paul S. Boyer
AMERICAN IMMIGRATION
    David A. Gerber
AMERICAN LEGAL HISTORY
    G. Edward White
AMERICAN POLITICAL
    HISTORY Donald Critchlow
AMERICAN POLITICAL PARTIES
    AND ELECTIONS L. Sandy Maisel
AMERICAN POLITICS
    Richard M. Valelly
THE AMERICAN PRESIDENCY
    Charles O. Jones
THE AMERICAN REVOLUTION
    Robert J. Allison
AMERICAN SLAVERY
    Heather Andrea Williams
THE AMERICAN WEST Stephen Aron
AMERICAN WOMEN'S HISTORY
    Susan Ware
ANAESTHESIA Aidan O'Donnell
ANARCHISM Colin Ward

ANCIENT ASSYRIA Karen Radner
ANCIENT EGYPT Ian Shaw
ANCIENT EGYPTIAN ART AND
    ARCHITECTURE Christina Riggs
ANCIENT GREECE Paul Cartledge
THE ANCIENT NEAR EAST
    Amanda H. Podany
ANCIENT PHILOSOPHY Julia Annas
ANCIENT WARFARE
    Harry Sidebottom
ANGELS David Albert Jones
ANGLICANISM Mark Chapman
THE ANGLO-SAXON AGE John Blair
THE ANIMAL KINGDOM
    Peter Holland
ANIMAL RIGHTS David DeGrazia
THE ANTARCTIC Klaus Dodds
ANTISEMITISM Steven Beller
ANXIETY Daniel Freeman and
    Jason Freeman
THE APOCRYPHAL GOSPELS
    Paul Foster
ARCHAEOLOGY Paul Bahn
ARCHITECTURE Andrew Ballantyne
ARISTOCRACY William Doyle
ARISTOTLE Jonathan Barnes
ART HISTORY Dana Arnold
ART THEORY Cynthia Freeland
ASTROBIOLOGY David C. Catling
ATHEISM Julian Baggini
AUGUSTINE Henry Chadwick
AUSTRALIA Kenneth Morgan
AUTISM Uta Frith
THE AVANT GARDE David Cottington

David A. Rothery

MOONS

A Very Short Introduction

OXFORD
UNIVERSITY PRESS

Great Clarendon Street, Oxford, OX2 6DP,
United Kingdom

Oxford University Press is a department of the University of Oxford.
It furthers the University's objective of excellence in research, scholarship,
and education by publishing worldwide. Oxford is a registered trade mark of
Oxford University Press in the UK and in certain other countries

First edition published in 2015

Impression: 1

Published in the United States of America by Oxford University Press
198 Madison Avenue, New York, NY 10016, United States of America

British Library Cataloguing in Publication Data
Data available

Library of Congress Control Number: 2015943217

ISBN 978-0-19-873527-4

Printed in Great Britain by
Ashford Colour Press Ltd, Gosport, Hampshire

Contents

List of illustrations

Moons

Chapter 1
The discovery and significance of moons

It would be meaningless to write about the discovery of our moon, the one that goes round the Earth—'the Moon', as I shall call it from now on because that's its name. The Moon is almost as obvious as the Sun. Clouds permitting, we can see it in the evening sky nearly half the time. If we are awake early, there it is in the pre-dawn sky most of the rest of the time. It can often be spotted in daylight too.

Humans have been well aware of the Moon for as far back as records go, but surely before then too because the Moon must often have been a welcome source of illumination at night. Possibly the oldest Moon-related artefacts are 30,000-year-old bone plates engraved with dots or lines, thought by some to be a way of keeping track of the Moon's phases, as it swells from new to full and then shrinks back from full to new on its 29.5-day cycle. The Moon's appearance changes in this way because its orbit carries it round the Earth, continually changing its position in the sky relative to the Sun so that the amount of the illuminated hemisphere we can see changes too.

It would be equally meaningless to try to pin down who first realized that the Moon goes round the Earth. To most ancient peoples it must have seemed pretty obvious that *all* the heavenly bodies go round the Earth. In fact that's wrong, and the Moon is actually the *only* natural body in the sky to go round the Earth. I'm not referring to the daily (24-hour) motion whereby the Sun and all other celestial bodies rise and set—that's just an apparent motion because the Earth is spinning—but to the Moon's 29.5-day journey round the sky relative to the Sun. This 29.5-day period is a combination of the 27.3 days that it takes the Moon to make a 360° circuit round the Earth plus just over two days more to compensate for the fact that the Earth has moved about a twelfth of the way round the Sun in the meantime.

In due course, the discovery of moons orbiting another planet showed that the Earth is not special so far as motion is concerned. This provided crucial evidence to help overturn deeply entrenched 16th- and 17th-century orthodoxy that held the Earth to be the hub of all creation.

Although some ancient Greek philosophers had preferred the notion that the Earth and planets move round a central fire (the Sun), they were in a minority. By the start of the 17th century, the long-established view of the universe had the Earth at the middle, with the known heavenly bodies going around it. The Moon was rightly believed to be closest, then Mercury, Venus, the Sun, Mars, Jupiter, and Saturn. Beyond this was a sphere bearing the stars. So far as most philosophers were concerned this sphere too rotated, because the Earth seemed motionless, although some Greeks and Indians had suggested that the Earth was spinning.

Greek philosophers and their later followers were wedded to the concept that heavens should be 'perfect', so they tried to interpret the observed movements of heavenly bodies in terms of uniform circular motion. As the precision of observations improved, more and more flaws appeared when trying to fit theory to observation. This led to the development of elaborate and cumbersome explanations, in which smaller circles were fitted on to larger circles in a system of 'epicycles', and also in which the speed of motion was uniform only when measured about a special point that did not coincide with the centre of the relevant circular track.

This elaborate Earth-centred (geocentric) view of the cosmos is generally referred to as the Ptolemaic system, after a Greco-Egyptian named Claudius Ptolemy who worked in Alexandria about AD 150, and was endorsed by the Catholic Church. In much of Europe it was dangerous to promote a contrary view, and geocentric ideas also held sway in China and across the Islamic world.

However, in the early 16th century the Polish astronomer Nicolaus Copernicus (1473–1543) developed a rival theory, which had the sphere of fixed stars surrounding the planets including the Earth that went round the Sun (a heliocentric model), and only the Moon going round the Earth. This is essentially correct (except that the stars are not fixed to a sphere—they are merely very distant), but to make it fit the available observations Copernicus had to invoke even more epicycles than were needed in the Ptolemaic system. It wasn't until Johannes Kepler (1571–1630) introduced elliptical, rather than circular, orbits round the Sun in 1609 that the heliocentric model became more elegant. It took until 1687 for Isaac Newton (1643–1727) with his laws of motion and theory of gravity to explain *why* orbits are ellipses.

Although Copernicus's model became known to many colleagues, he was reluctant to publish. It was only in the year of his death, 1543, that his great work *De revolutionibus orbium coelestium* (*On the Revolutions of the Heavenly Spheres*) was published. Contrary to popular myth, this was not immediately banned by the Church, but its views were certainly controversial because they contradicted the biblical view of the cosmos with the Earth as the centre of creation.

Enter Galileo Galilei (1564–1642), an Italian scientist working in Padua. In 1610 he turned one of the world's first telescopes (which he had built himself) skywards. As well as discovering the phases

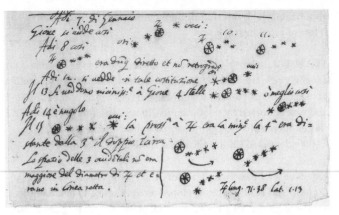

1. Part of a page of notes by Galileo, documenting four moons of Jupiter on successive nights from 7 to 15 January, 1610 (except 14 January, which was cloudy).

of Venus and a multitude of faint and otherwise invisible stars in the Milky Way, he saw up to four faint 'stars' accompanying Jupiter (Figure 1). He was able to note that these shifted back and forth to either side of Jupiter, and concluded after only a few nights' observations that these objects were orbiting Jupiter in a plane very close to his line of sight.

Galileo disseminated his observations in March 1610 in a pamphlet that he called *Sidereus Nuncius* (*Starry Messenger* or *Starry Message*). That autumn, when Jupiter reappeared in the sky, several astronomers, including the Englishman Thomas Harriot (1560–1621) and the German Simon Marius (1573–1625), confirmed the existence of these four moons of Jupiter. In fact Marius claimed that he had discovered them himself in 1609.

They were described as 'satellites' (still a correct alternative to 'moons') before they were described as 'moons'. Galileo himself used neither term (at least at first), referring to them as 'stars', though we can take this as a reference to their point-like appearance rather than an indication that he thought they were the same as conventional stars. Marius called them 'Jovian planets' in his 1614 work *Mundus Jovialis* and Harriot also wrote of his own experience of seeing 'the new planets'. It was Kepler who first called them satellites, when writing of them in 1610, adopting a Latin word meaning 'one who attends a more important person'.

Irrespective of whether there was any truth in Marius's claim, Galileo was the first to publish, and so is credited with the first observational proof that all celestial motion is not centred on the Earth, nor indeed all on the Sun.

Making use of moons

Jupiter's moons orbit their planet in a predictable, clockwork fashion. However, within fifty years of their discovery the Italian Giovanni Cassini (1625–1712), working in Paris, had noted some slight but systematic discrepancies in the regularity with which the moons disappeared into Jupiter's shadow. The intervals between disappearances (eclipses) become slightly shorter when the Earth's orbit is bringing it closer to Jupiter, and lengthen slightly as the two planets draw apart. Cassini rightly suggested that this is because light does not travel infinitely fast, as was then believed, but in fact takes a measurable time to get from Jupiter to the Earth. In 1676 he argued that it takes about ten or eleven minutes for light to cross the Earth–Sun distance, which was pretty close to the correct value of eight minutes and thirty-two seconds. His assistant Ole Rømer (1644–1710) soon came up with observations that enabled a more precise estimate, but it was about fifty years before this moon-based demonstration of the finite speed of light became generally accepted. It also needed a good measurement of the Earth–Sun distance before this speed could be confidently converted into familiar units, though in fact Cassini had made an estimate of the Earth–Sun distance in 1672 that was only about 7 per cent below the correct value of 150 million km.

Telescopes had to improve beyond the rather crude devices built by Galileo and his contemporaries before other moons could be discovered. The Dutchman Christiaan Huygens (1629–95) built a better telescope, and in March 1655 he found the first known moon of Saturn, referring to it as *Luna Saturni* (Saturn's moon) and in so doing probably became the first to use the term 'moon' to refer to the satellite of another planet. Cassini found another four fainter (and therefore smaller) moons of Saturn between the years 1671 and 1686.

To some at the time there seemed to be a logic in Saturn having five moons whereas Jupiter had only four. However, many more discoveries were to come and these destroyed any semblance of a simple pattern. In 1781, the planet Uranus was discovered by William Herschel (1738–1822), a Hanoverian working in England, and in 1787 he discovered two of its moons followed by two more moons of Saturn in 1789. The next moon to be seen was the largest moon of Neptune, found by the Englishman William Lassell (1799–1880) just seventeen days after the planet itself had been discovered in 1846. Lassell is jointly credited with the American father and son William (1789–1859) and George (1825–65) Bond for the discovery of an eighth moon of Saturn in 1848, and individually discovered two more moons of Uranus in 1851.

The two tiny moons of Mars (the only ones it has) were discovered in 1877 by the American Asaph Hall (1829–1907). A fifth moon of Jupiter was discovered by the American Edward Emerson Barnard (1857–1923) in 1892, nearly 200 years after Galileo had found the first four. The delay was because this one is much smaller and also closer to Jupiter, which are both factors that make it hard to see. It was the last moon to be discovered visually. Subsequent discoveries were by photography, the first being a ninth moon of Saturn found by the American William Pickering (1858–1938) in 1899 on plates that had been taken the previous year in Peru. More recently discoveries have been achieved by digital cameras fitted to telescopes or on spacecraft.

So, there are lots of moons in our Solar System—in fact far more than I have yet mentioned—but is knowing about them of any use? Perhaps surprisingly, the moons discovered by Galileo found a practical use even in the 17th century. Until chronometers (clocks) had been developed that could keep accurate time despite being transported (achieved in the late 18th century), it was immensely difficult to determine longitude. On land you could try to measure the distance between places, whereas at sea the best you could do was to estimate distance travelled from course and speed. This sometimes went disastrously wrong, as in 1707 when a homeward-bound British naval fleet struck rocks off the Isles of Scilly with the loss of more than 1,400 lives.

The independent clockwork of Jupiter's moons offered a partial solution, especially after Cassini produced a set of tables accurately predicting the times of their eclipses. All you had to do to fix your longitude was to measure the time between a particular eclipse and local noon at your location, and then use Cassini's tables to decide how far east or west you were from Cassini's reference longitude.

Cassini himself secured a commission from Louis XIV of France to improve the maps of his kingdom, and in the 1670s he and his team set about determining the longitudes of the major cities of France, relative to Paris. They would observe an eclipse of one of Jupiter's moons and then use a pendulum clock (a perfectly fine chronometer if we don't try to move it) to measure how long this occurred before local noon (determined by observing when the Sun reached its peak altitude).

Sometimes the truth is hard to bear. Many cities were found to be up to 100 km closer to Paris than expected, showing France's east–west extent to be less than previously believed. Louis reputedly complained that his own astronomers had deprived him of more territory than his enemies ever had.

Cassini successfully exported his technique overseas. The difficulties in keeping a telescope trained on Jupiter from a moving ship made it impractical while actually at sea, but it was fine for determining longitude ashore, and indeed was still being used in the early 19th century as an aid to mapping in the American West.

The Moon itself provided another means of determining longitude, because it moves across the sky by an amount roughly equal to its own diameter every hour. By measuring the position of the Moon relative to a known star we can determine the time, but we have to correct for parallax because the Moon is sufficiently close to the Earth for its position in the sky relative to the background stars to be slightly different when seen from different locations. The principles of the method were proposed by the Englishman Edmond Halley (1656–1742), later of comet fame, around 1683. The necessary measurements required precise determination of the angle between the Moon and a star followed by difficult calculations that could take over half an hour to perform, until some tables were developed that reduced this to about ten minutes. This method of using the angular separation between the Moon and reference stars was known as 'lunars' and was widely used at sea between 1767 and about 1850, by which time reliable marine chronometers had become affordable. However, it continued to be taught to navigators as a back-up technique for at least a further fifty years.

Another use of moons is to measure the mass of the object about which they orbit (or rather the combined mass of the two objects). This is because the square of the orbital period is proportional to the cube of the orbital radius divided by the sum of the masses. If the moon is much smaller than the planet, then its own mass is negligible, so the orbital period tells us the planet's mass. To turn this measurement into familiar units, we need to divide the number we get for the mass by a number that is related to the 'gravitational constant' G (which basically tells us how much mass it takes to produce a given strength of gravity).

Had G been known early enough, then scientists could have used the Moon's orbital period to weigh the Earth. It didn't work out that way, because G was first adequately determined in 1798 by the Englishman Henry Cavendish (1731–1810) in an experiment that essentially determined G and the mass of the Earth at the same time. However, once G was known and the scale of the Solar System had been determined (Cassini had not been far out, and the uncertainty had been considerably reduced a hundred years later thanks to observations of transits of Venus across the Sun in 1761 and 1769), the orbital periods of other planets' moons could be used to determine the planets' masses and their densities. This revealed, for example, the enormous masses of Jupiter and Saturn (318 and 95 times the Earth's mass, respectively) and also their low densities (Saturn has only 69 per cent the density of water).

Venus and Mercury have no moons, and so their masses had to be estimated from the tiny perturbations that they cause to the Earth's orbit, or to a conveniently passing comet. This left significant uncertainties until spacecraft visited them and experienced their gravitational pulls at close quarters.

The naming of moons

When moons began to be discovered there was no system for naming them. Galileo gave his four moons of Jupiter the collective name *Medicea Sidera* (the Medician Stars) after the family of his patron Cosimo de' Medici. He distinguished them individually by Roman numerals I, II, III, and IV, working outwards from Jupiter. In 1614 Simon Marius proposed the names Io, Europa, Ganymede, and Callisto, after lovers of the god Zeus (the Greek equivalent of Jupiter). Galileo, perhaps aggrieved by Marius's rival claim to discovery, would have nothing to do with those names. They did eventually become the officially recognized names that we use today, though a system of Roman numerals continues in parallel use (for newer discoveries as well). However, Galileo is honoured too, because as a group the four moons that he discovered (which are much larger than any of Jupiter's other moons) are called the 'Galilean moons'.

Cassini referred to the four moons of Saturn that he discovered as *Sidera Lodoicea* (the Louisian Stars) after his patron Louis XIV of France, alongside the larger *Luna Saturni* that Huygens had discovered. Other astronomers preferred a Galileo-style Roman numeral convention to identify individual moons. However, the practice of numbering outwards from the planet was a recipe for confusion, because new discoveries could change the order. For example, Huygen's *Luna Saturni* successively bore the numbers II, IV, and then V. By international consensus the numbering system was frozen after William Herschel's 1789 discoveries, so that the designations of individual bodies would never again be reshuffled. In 1847 William's son, John Herschel (1792–1871), suggested a coherent set of names for all seven of the then-known moons of Saturn, which we still use today. He called the largest one Titan after the collective name for the mythological siblings of Chronos (the Greek Saturn), and named the others after some of the individual Titans: Iapetus, Rhea, Tethys, Dione, Enceladus, and Mimas. When Lassell co-discovered the next moon of Saturn in 1848, he named it Hyperion, after another of the Titans in accordance with John Herschel's theme.

After he discovered the third and fourth moons of Uranus, Lassell invited John Herschel to name them along with his father's two previous discoveries. He chose Titania and Oberon (after the faerie queen and king in Shakespeare's *A Midsummer Night's Dream*), Ariel (after a sky spirit in Alexander Pope's *The Rape of the Lock* who also appears in Shakespeare's *The Tempest*), and Umbriel (a melancholy spirit in *The Rape of the Lock*).

Surprisingly, Lassell did not name the moon of Neptune that he discovered in 1846. Its present name, Triton (son of Poseidon, the Greek Neptune), was suggested as late as 1880.

The names of Mars' two tiny moons, Phobos and Deimos, were chosen by their discoverer based on a suggestion by Henry Madan (1838–1901), a science master at Eton school, from Book XV of *The Iliad* in which Ares (the Greek Mars) summons the twin brothers Fear (Phobos) and Terror (Deimos).

The names of subsequently discovered moons (there have been no more at Mars) have followed the theme that became established at each planet when the first names became generally accepted. Since its foundation in 1919, a body known as the International Astronomical Union (IAU) has been the arbiter of names (and their spelling) for objects in the Solar System, and also their surface features. A newly discovered moon is awarded a temporary designation (for example S/2011 J2 is the second satellite of Jupiter to have been discovered in 2011), and receives formal name only when its orbit has been well characterized. Names for Jupiter's moons are taken from lovers and descendants of Zeus/Jupiter. It is fortunate that mythology provides plenty to choose from, because Jupiter has fifty named moons and sixteen still awaiting formal naming.

Saturn has a similar number of known moons, of which fifty-three have been named. The first eighteen were named after Titans and their offspring. John Herschel's original theme was then broadened to include other giants from Greco-Roman, Gallic, Inuit, and Norse mythologies, allocated according to their orbits.

Except for Belinda, which is another name deriving from Pope's *The Rape of the Lock*, the moons of Uranus continue to receive names of (mostly female) Shakespearean characters. These range from the exotic Sycorax (a witch who was mother to Caliban in *The Tempest*) to the prosaic Margaret (a maidservant in *Much Ado About Nothing*). Neptune's moons are allocated names of mythological characters associated with Poseidon/Neptune.

How many moons?

It is hard to keep track of how many moons are known in our Solar System. Since the year 2000 there have been scores of discoveries. These were made mainly using ground-based telescopes and the Hubble Space Telescope, but NASA's Cassini spacecraft in orbit about Saturn found several additional small moons of that planet. The bodies known to possess moons include all the planets from Earth out to Neptune. It seems certain that neither of the inner planets, Venus and Mercury, can have a moon bigger than a kilometre or so in size, or it would have been discovered by telescopes or by orbiting spacecraft.

The 190 known moons of our Solar System's planets are listed in tables in the Appendix. There I quote 'mean radius' mainly because the smaller moons, especially those less than about 200 km in radius, can be markedly non-spherical in shape. For comparison, the Earth's mean radius is 6,371 km. I list density in thousands of kilograms per cubic metre. This is equivalent to tonnes per cubic metre, or grammes per cubic centimetre. For comparison, using the same units the density of water is 1.0.

Various icy bodies beyond Neptune have moons too. Among those, Pluto has the largest known family, and this is included in a table in the Appendix. Pluto-like bodies are substantial in size, and it is perhaps no great surprise that some of them should have moons. However, several asteroids, including one example less than a kilometre across, have been found to have moons too (which by definition are even smaller than the object that they orbit). Few people expected that, and it is a topic to which I will return in Chapter 7.

Does the Earth have more than one moon?

You may come upon claims on the Internet or in quizzes that the Earth has one or two small moons in addition to the Moon. These are misguided. Although there are very many artificial satellites in space, no small natural objects have been found in permanent orbit around the Earth. Any that there are must be less than about ten metres across or they would have already been discovered, and calculations suggest that such 'minimoons' could not have stable orbits.

However, there are small asteroids whose orbits cross that of the Earth in a way that enables them to become captured temporarily, during which interval they are indeed minimoons. An example is the five-metre-wide object 2006 RH_{120}. This was discovered telescopically in 2006, and the following year it made four loops round the Earth, each one different in shape. Its path was mostly well beyond the orbit of the Moon, but its closest approach brought it to about 70 per cent of the Moon's distance from us. It broke free after eleven months as a temporary satellite of the Earth, its orbit having been complex and unstable partly because it felt the Moon's gravitational pull as well as the Earth's. By 2017 it will be on the far side of the Sun, but will pass close again in 2028, offering another opportunity for temporary capture. What will happen can't be predicted, because irregular-shaped objects like this are small enough for solar radiation pressure to affect their trajectories. We can't estimate the size of the radiation-driven perturbations, because we know neither the object's shape nor its density.

Actually, 2006 RH$_{120}$ may not be a natural object; it could be the third stage of one of the Apollo rockets used in NASA's 1969–72 Moon programme. Irrespective of this, some studies suggest that at any one time the Earth may be accompanied by up to a few dozen temporarily captured orbiters less than two metres in size that arrive, complete at least one loop, and are then lost. Larger examples are rarer, and it is probably only once in 100,000 years that something in the hundred-metre size range becomes temporarily captured in this way.

Another class of objects sometimes misreported as 'moons' of the Earth are asteroids whose average orbital period about the Sun is exactly the same as the Earth's. Such an asteroid has a path round the Sun that lies very close to the Earth's orbit, but the Earth's gravity exerts a strong influence on it, so that it cycles through a repeating pattern as follows. Imagine the asteroid following an orbit slightly inside the Earth's orbit. This means that the asteroid will be progressing round the Sun slightly faster than the Earth, and eventually it will catch up with it. As it draws near, the Earth's gravity pulls the asteroid into a larger orbit, which takes longer to complete so that the asteroid is now travelling more slowly round the Sun and lags behind the Earth. Eventually, after they have both made several orbits, the Earth almost catches up with the asteroid and pulls it into a smaller orbit. Now travelling faster than the Earth, the asteroid draws ahead and we are back to where we started.

From the perspective of the Earth, the asteroid's path resembles a horseshoe (with the Earth in the incomplete part of the ring). Because of that, these orbits are sometimes described as 'horseshoe orbits', but it is important to realize that the asteroid does not orbit the Earth and is always travelling in a forward direction around the Sun. At least three such asteroids are known, of which the largest, 2010 SO$_{16}$, has a mean radius of about 300 metres. A few other asteroids have less circular orbits with periods matching the Earth's, which leads them to migrate around a mean position either ahead of or behind the Earth, without migrating around a complete 'horseshoe'. The best known of those is 3753 Cruithne, which has a mean radius of about 2.5 km. This too orbits the Sun, not the Earth, and is not a moon. You can find more about the strange orbits of such 'quasi-moons' at a link in Further reading.

Can moons have moons?

Planets go round the Sun, and moons go round their planets, so a natural question is whether any moon could have a natural satellite of its own—a moon of a moon. In the case of our own Moon, it has had several temporary moons in the form of artificial satellites that humans have placed there in recent years. However, these have all been in unstable orbits leading them to crash on to the lunar surface after a few years. This is because the Moon's gravity is insufficient to dominate the region of space that surrounds it, because the much more massive Earth is too close.

The Moon has a stable orbit round the Earth because the Earth's gravity is strong enough to outcompete the Sun's gravity to a distance of about one million km, a volume of space that is described as the Earth's 'Hill sphere', after the American astronomer George William Hill (1838–1914) who defined the concept. The Moon's orbit lies well inside this, and so enjoys long-term stability. The Moon's own Hill sphere is 60,000 km across, but any object orbiting the Moon even within its Hill sphere experiences a sufficient pull from the Earth's gravity to cause its orbit to shrink over time. It turns out that the same is true for moons of other bodies too, so that orbits about moons anywhere in the Solar System are not stable in the long term. The duration over which an orbit about a moon could persist varies from years to many millions of years, according to a complex balance of forces, but is much shorter than the 4.5 billion-year age of the Solar System.

Thus there are no moons of moons, and if one were to be discovered it would almost certainly be a short-lived situation.

Chapter 2
The Moon

Now for some individual moons in detail. I will begin with the Moon, because this is the one about which most is known, and of course you will have seen it for yourself.

The Moon has been referred to by that name as far back as can be traced in Germanic languages (Old English *Mōna*, proto-Germanic *Mǣnōn*). In Latin it was *Luna* (whence the English adjective *lunar*, French *Lune*, Spanish/Italian *Luna*, and the identically pronounced Russian Луна). In ancient Greek the Moon was Σελήνη (*Selene*) from which we get the prefix used in terms such as 'selenographic coordinates' (the latitude/longitude system used for mapping the Moon). The term 'moon' was eventually transferred by analogy to any object orbiting another planet, though as we have seen the term satellite was invoked for them by Kepler much more quickly.

The Moon is a substantial body. Only four other moons are bigger (and also more massive, in the sense of having a greater mass): Jupiter's Io, Ganymede, and Callisto, and Saturn's Titan. If the Moon were to orbit the Sun independently there is no doubt that it would be ranked among the 'terrestrial planets', which are the Sun's four innermost planets. Table 1 shows both the equatorial and the polar radii of these bodies, because their rotation distorts their shapes so that they bulge at their equators and are flattened

Table 1 The Moon and the terrestrial planets compared

Name	Equatorial radius/km	Polar radius/km	Mass/ 10^{24} kg	Density/ 10^3 kg m^{-3}
The Moon	1,738.1	1,736.0	0.07342	3.344
The Earth	6,378.1	6,356.8	5.9726	5.514
Mercury	2,439.7	2,437.2	0.3301	5.427
Venus	6,051.8	6,051.8	4.8676	5.243
Mars	3,396.2	3,376.2	0.6417	3.933

towards their poles. This makes the equatorial radius greater than the polar radius. Venus is an exception: it spins nearly 250 times slower than the Earth and shows no measurable flattening.

Although not much smaller than Mercury, the Moon contains considerably less mass because its overall density is lower. The reason for this is that Mercury has a very large, dense iron-rich core surrounded by a relatively thin rocky mantle and crust, whereas the Moon is made almost entirely of rock. Its core, if it even has one, is no more than about 300 km in radius.

Mercury has a thick molten zone in the outer part of its core. This is circulating, and because it is an electrical conductor it acts like a dynamo and generates a magnetic field that encompasses the planet, partially shielding its surface from bombardment by the charged particles (cosmic rays) that stream out from the Sun. The Moon lacks such a magnetic field, and its surface is exposed to whatever the Sun throws at it.

Like Mercury, the Moon's own gravity is too slight for it to retain the gases necessary for an atmosphere, but there are some gaseous atoms above the surface. For example, sodium and potassium atoms are released by 'sputtering' when the surface is hit by the solar wind, helium is added directly by the solar wind, and argon escapes from the lunar interior where it is produced by radioactive decay of an isotope of potassium. The total 'atmospheric pressure' of these atoms is about 3×10^{-15} (3 million billionths!) of the Earth's surface atmospheric pressure. This is so low that atoms are more likely to escape into space than to bump into each other, making the whole of the Moon's atmosphere equivalent to the very tenuous outermost zone of the Earth's atmosphere, which is known as the 'exosphere'.

The virtual absence of an atmosphere leads to a very large day–night range of temperature at the Moon's surface, which swings from about 120°C near the equator at noon to around –150°C at night. These day–night variations do not penetrate deeply into the lunar soil or 'regolith', and at about a metre below the surface the temperature is believed to be a fairly constant –35°C. There are some craters near the poles whose floors are never illuminated by the Sun, and where the surface temperature is permanently below about –170°C.

Phases, orbit, and rotation

The Moon's appearance is familiar to most people. Even with the naked eye we can make out dark patches on its surface. The first telescopes enabled observers such as Galileo and Harriot (who did the better job) to map these dark patches and also make out smaller features such as craters and mountains. If you look with a pair of binoculars, you will probably be able to see more than they did.

Whenever you look, you will see the same hemisphere of the Moon as the one drawn by Harriot and Galileo, because the same side of the Moon always faces the Earth. What does vary is how much of it we can see lit by the Sun at any particular time, depending on where the Moon is in its 29.5-day orbit. When the Moon lies between the Earth and the Sun (which we call new Moon) we can't see it at all. It rarely comes *exactly* between the Earth and the Sun, causing a solar eclipse, because the Moon's orbit round the Earth is tilted at about 5° relative to the Earth's orbit round the Sun, with the result that the Moon usually passes (invisibly) either a little above or below the Sun rather than across its face.

A couple of days after new Moon, the Moon has drawn far enough away from the Sun in the sky for it to become visible as a thin crescent, which grows until half the Earth-facing hemisphere is lit. Somewhat confusingly, this is called 'first quarter', which refers to a quarter of the 29.5-day cycle being completed rather than to how much of the disc is illuminated. The visible illuminated area continues to grow (its shape is now referred to as 'gibbous'), until, nearly fifteen days after new Moon, the Moon is on the far side of the Earth from the Sun. The Earth-facing hemisphere is then fully lit, and we call this 'full Moon' (though if the Earth gets *exactly* in the way it stops direct sunlight reaching the Moon, and there is a lunar eclipse). As the Moon continues in its orbit, the illuminated fraction shrinks until only half the Earth-facing hemisphere is lit (third quarter), then the shape passes through a waning crescent until it disappears close to the date of the next new Moon.

If you think that seeing the same side all the time means that the Moon doesn't rotate, think again. To keep the same side facing the Earth, the Moon has to rotate exactly once per orbit. This is demonstrated in an animation to which I have put a link in the Further reading. Almost every known moon is in such a state of 'synchronous rotation', because of tidal drag that forces its rotation to keep pace with its orbit.

This comes about because the mutual gravitational attraction between two nearby bodies, such as a planet and its moon, distorts their shapes slightly. A tidal bulge (about a hundred metres in the case of the Moon) is raised in the middle of each facing hemisphere. This is because the nearside is closer to the other body than its centre is, and so experiences a slightly stronger gravitational pull. There is an equal bulge on the farside because the body's centre is pulled more strongly towards its neighbour than its farside is.

If a moon's axial rotation were faster than its orbital period, its shape would have to distort continuously as the tidal bulge migrated around the globe so as to stay lined up with the planet. This would use up energy until the moon's spin had been slowed down to match the orbital period. So although the Moon and most moons of other planets were probably spinning faster to begin with, their spin now matches their orbital periods.

Incidentally, although the Moon has a nearside and a farside, it doesn't have a permanent dark side. The 'dark side of the Moon' is an acceptable metaphor for the hidden side about which we once knew nothing. However, because the Moon rotates all sides see the Sun over the course of a single orbit—though of course the Sun can illuminate only half the Moon at any one time.

Although the Moon is orbiting the Earth, it is at the same time accompanying the Earth in its year-long journey round the Sun. The Earth's speed round the Sun is about 30 km/s, which is much faster than the Moon's orbital speed round the Earth: 1.022 km/s when the Moon is at its closest (a position known as perigee) and 1.076 km/s when at its furthest (apogee). Consequently, the Moon's path weaves to either side of the Earth's orbit, but it is always convex outwards from the Sun. It never backtracks.

The face of the Moon

We are the first generations to know what the Moon's farside looks like. This remained unknown until October 1959 when the Soviet probe Luna 3 was sent out beyond the Moon and beamed back the first blurry pictures. A much clearer modern view of the farside is included in Figure 2, which shows the Moon from four different directions. The nearside view shows what we can see from Earth

2. Four views of the Moon assembled from images acquired by NASA's Lunar Reconnaissance Orbiter. The nearside (centred at 0° longitude), the farside (centred at 180° longitude), and two intermediate views (east, centred at 90° longitude and west, centred at 270° longitude).

at full Moon. The farside view shows the opposite face of the Moon, seen when that side is fully illuminated by the Sun.

There is a striking contrast between the nearside and farside. About half the nearside is occupied by dark patches. In the 17th century, astronomers mistook these for seas (or at least the dry beds of former seas), and denoted them by the Latin word for sea, *mare*, of which the plural is *maria*. We now know that these areas have in fact been flooded by vast outpourings of volcanic lava resembling basalt. They have never been occupied by water, but *mare* and *maria* are still used today, both to refer to them in general and in their formal (IAU-approved) names. The correct pronunciations for *mare* and *maria* are 'MAH-ray' (not as in a female horse) and 'MAH-ria' (not as the girl's name).

On the farside only a few, relatively small, areas are occupied by maria. This is believed to be partially due to the crust being thicker on the farside, so that it has been harder for magma to reach the surface. It also suggests that the Moon has been in synchronous rotation with the same hemisphere facing the Earth since at least 3.8 billion years ago, when the majority of maria began to fill with lava.

The brighter surface outside of the maria is a different sort of rock. It is largely made of a kind of feldspar mineral called anorthite, giving the name 'anorthosite' to the rock. It represents the Moon's oldest crust, up to 4.5 billion years old. This terrain is referred to as the lunar highlands. Low-lying regions of 'highlands' on the nearside (most of them large basins excavated by impacts) were flooded by mare basalts, but this didn't happen to the same extent on the farside.

From Earth, we are able to see glimpses of alternate edges of the farside, because the Moon's orbit is not quite circular but an ellipse, which brings the Moon's centre to a distance of only 363,300 km from the Earth's centre at perigee but 405,500 km at apogee. Orbital speed is faster when closer and slower when further away, but the Moon's spin maintains a constant rate. As a result, around apogee the spin outpaces the orbital motion so we can see a little way round the average leading edge of the Moon, whereas the terrain close to the average trailing edge is rotated out of view. As perigee approaches, we see a little way beyond the average trailing edge instead. In total about 59 per cent of the Moon can be seen from Earth at one time or another, though we only ever get a foreshortened, highly oblique, view of the terrain near the edges.

This side-to-side wobbling of the Moon's face is a phenomenon called 'libration', and you can see it in action if you follow another of the links the Further reading. This also shows how the Moon looks slightly bigger (because it is closer) at perigee, and how the visible shape of the illuminated half of the Moon (the Moon's 'phases'), as seen from the Earth, changes as the Moon progresses round its orbit.

The images in Figure 2 lack any shadows, because the viewing geometry is directly down-Sun. This exaggerates the contrast in

3. The Moon as seen from Earth about three days eight hours before full Moon. The terminator is on the left. The illuminated edge of the visible disc (on the right) is referred to as the 'limb'.

the fraction of sunlight that different surfaces reflect (a property called 'albedo'), but topography is not apparent. Figure 3 shows a view of the gibbous Moon, in which the shadows are increasingly more obvious the closer you look to the terminator, which is the line dividing night from day.

In particular, many circular holes are apparent near the terminator thanks to shadows inside their rims. These are craters, whose origin I will discuss next, but first look at how their appearance differs between Figure 3 and the nearside view in Figure 2. The most obvious crater in the upper left of Figure 3 occurs on a dark background, because it lies in a region of mare. This is a 93 km diameter crater named Copernicus, thought to have formed about 0.8–1.1 billion years ago. It stands out strongly in the Figure 2 nearside view, because its floor penetrates through the low-albedo mare basalts into the buried high-albedo highland crust.

A prominent crater in the lower left of Figure 3 is in the middle of a bright patch in Figure 2. This is an 86 km diameter crater named Tycho. It has a brighter (higher albedo) floor than Copernicus because there is no basalt involved and it is all freshly pulverized highland crust. Some almost equally bright ejecta thrown out of the crater covers the surface surrounding it, and beyond this there are bright 'rays' radiating away, which are streaks of finely powdered ejecta. Apollo 17 landed about 2,000 km north-east of Tycho, but the astronauts collected fragments from one of Tycho's rays that crossed the landing site. Analyses carried out back on Earth found these to be about 100 million years old, showing that Tycho is much younger than Copernicus.

You can see several other rayed craters in the four images in Figure 2. Rays are prominent only when the Sun is close to overhead. They fade with age and are associated only with the youngest craters. Copernicus has rays, but they are less prominent than Tycho's.

Impact craters

Craters are circular rimmed depressions that dominate the Moon's surface. In fact they are abundant on the surfaces of most solid bodies in the Solar System, including almost every moon. They are rare only where ancient surface features have been erased by erosion, or buried by something such as lava flows or wind-blown dust.

In the case of the Earth, crust is recycled because of the action of plate tectonics, which crumples the edges of continents during collisions and pushes old ocean floor down into the mantle. This adds to the effects of general erosion and burial, so that the Earth has very few obvious Moon-like craters. Although nearly 200 have now been documented, few are spectacular, most are degraded, and none were well known when people began to notice and speculate about lunar craters.

It was Galileo who first referred to them as 'craters', using a Greek word meaning a drinking vessel. Their presence on the Moon undermined the notion that the Moon, being a heavenly body, should have a flawlessly smooth surface—though surely that was already hard to justify, given the blemishes represented by the obvious dark patches that can be seen even without a telescope, and which we now know as the maria.

The origin of lunar craters was a matter of controversy for some centuries. Most scientists regarded them as some kind of volcanic phenomenon, such as holes made by explosive eruptions or simply the scars left by vast bursting bubbles. The most common alternative theory—that they are the scars left by external objects hitting the Moon—was less popular for most of the time. Critics of that theory pointed to the fact that, with very few exceptions, the Moon's craters are circular in outline whereas impactors should arrive at all angles so that we would expect a large proportion of asymmetric or elongated craters. Others countered that lunar craters are very unlike what we see associated with most terrestrial volcanoes, so that a volcanic origin did not fit with the observations either.

It may seem obvious now that there are numerous bits of rock hurtling through the Solar System at high speed, but even as recently as 1808, US President Thomas Jefferson (by no means as ignorant of science as some of his successors) was publically sceptical. Of the results of inquiry by representatives of Yale University into a meteorite that was seen to fall and then collected in Connecticut, he is reputed to have declared that he 'would rather believe that two Yankee professors would lie than believe that stones fall from heaven'.

Evidence swung firmly in favour of an impact explanation for the Moon's craters in the 1960s. Three things helped with this. First, the American Gene Shoemaker (1928–97) studied the 1.2 km diameter Barringer Crater (also known as Meteor Crater) in Arizona, and found that the mineral quartz had been converted to denser forms of silica that form only under high pressure. He found identical mineralogic changes at underground nuclear test sites, and rightly concluded that the Barringer Crater had been excavated by high-pressure shock waves generated when a projectile struck the ground at tens of km per second (which is the sort of relative velocity expected when orbits intersect). Second, close-up images of the Moon's surface sent back by unmanned probes revealed that it has craters down to the smallest size visible. This could not be reconciled with any kind of volcanic phenomenon, whereas it posed no problem for impacts by objects hitting an airless body. Third, laboratory experiments resulted in convincing-looking craters with circular outlines when a 'hypervelocity' projectile was fired into a target in a vacuum chamber.

Figure 4 is a series of three nested and progressively higher-resolution (i.e. more detailed) views of lunar craters,

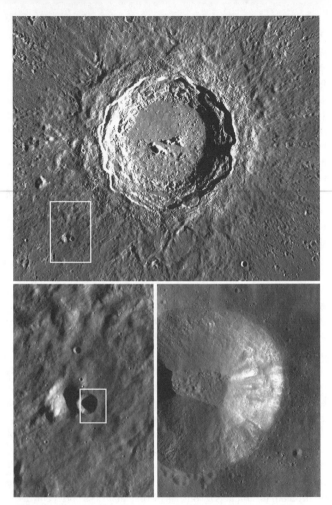

4. Copernicus crater and region seen by the Lunar Reconnaissance Orbiter at different resolutions. Top, Copernicus itself, seen at 250 metres per pixel. The image is about 200 km across, and the Sun was low in the east. The box outlines the higher-resolution view at lower left. Lower left, 20 km wide area seen at sixty-four metres per pixel. The box indicates the higher-resolution view to the right, the 7 km crater Copernicus B is immediately to the west of this box. Lower right, eight metres per pixel view of an unnamed 3.3 km wide crater, recorded with the Sun higher in the sky, and to the west.

beginning with Copernicus itself. Characteristics to note are that the floor of Copernicus is flat, except for a cluster of peaks near its centre. The crater floor is 2.5 km below the level of the plains surrounding the crater, and is bordered by a raised rim that is 1 to 1.5 km above the plains, giving a maximum drop from rim to floor of about 4 km. The central peaks rise about 1 km above the crater floor, so their summits are a long way below the crater rim and in fact do not rise even to the level of the surrounding plains.

The terrain outside Copernicus to a distance of more than one crater-radius is mantled by ejecta thrown out when the crater was formed. This buries any older craters, but the further away you look, the more smaller craters you can see. There are two near the middle of the medium-resolution view in Figure 4. The one to the left is about 7 km across, and is named 'Copernicus B'. It has a subdued shape as if its outlines have been smoothed by burial below Copernicus ejecta. An unnamed 3.5 km crater overlaps it, and has a much sharper appearance. This is younger, and was formed either by a 'secondary impact' when a large lump of Copernicus ejecta stuck the ground or at some later time by an unrelated small asteroid strike.

Neither Copernicus B nor its neighbour has peaks in its centre, and this illustrates an important observation: on the Moon central peaks are absent in craters less than about 10–20 km in size. Craters larger than that usually have a single central peak, unless they approach Copernicus in size when the single peak is replaced by a cluster of peaks. At even larger sizes—more than about 350 km across—crater structure becomes a double or even a multiple ring.

The highest-resolution view in Figure 4 was recorded with pixels only 8m across, and includes most of the younger crater that overlaps Copernicus B, which is about 300 metres deep. It was recorded with the Sun in the west and rather higher in the sky than for the two other views, so that the crater floor is free of shadows. At this level of detail you can see that the crater's inner slopes lack terraces of the kind so well seen in Copernicus. Outside the crater you can make out boulders as small as about twenty metres across and craters as small as about fifty metres.

There is no longer any doubt that the vast majority of craters on the Moon are a result of impacts. Asteroids, which are rocky or iron-rich bodies, typically strike the Moon at about 17 km/s. Comets, which are mainly icy bodies hurtling in from the outer Solar System, tend to strike faster at about 50 km/s.

As soon as an impactor hits the Moon's surface at speeds like this, shock waves radiate from the point of impact, which is why craters are circular and the angle of impact scarcely matters. The shockwaves melt and fragment both the target material and the impactor itself, and fling shattered and melted material outwards as ejecta. The first ejecta to be flung out travels the fastest, and goes furthest. During excavation, the crater gets wider rather than deeper. As the energy of the impact becomes expended the last ejecta barely manages to flop over the edge, contributing to the raised rim. The later ejecta falling close to the crater comes from deeper than the earlier ejecta, so any layering in the target material is inverted in the ejecta deposit.

Central peaks form, in sufficiently large craters, by a process called 'elastic rebound'. You can see similar rebound in high-speed videos of water droplets hitting the surface of a pond, except that in impact cratering the peak becomes frozen in place before it has chance to subside.

Craters end up thirty to a hundred times wider than the projectile that caused them, and the whole process is very fast. At typical impact speeds, to make a 93 km crater such as Copernicus would require a rocky asteroid about 5 km in diameter or an icy comet about 4 km in diameter, and would take about a couple of minutes to accomplish. The inner walls of craters as large as this tend to be unstable, and they subsequently collapse into a series of concentric terraces. To make a 7 km crater like Copernicus B would require a 200-metre asteroid or a hundred-metre comet, and would be all over in about thirty seconds.

Fortunately for us, because they would hit the Earth even more often than the Moon, impactors capable of making Copernicus-sized craters are now very rare. However, during the first billion years of the Solar System's history there was much more debris around, and some of it was big enough to make craters thousands of kilometres across (usually referred to as impact basins). There is no trace of these on the Earth any more, but many survive on the Moon.

On the nearside, these giant craters subsequently became flooded by lavas. Circular outlines attest to their origin, though some nearside basins have become sufficiently overfilled by lava that the flooded regions have merged. In the upper left of the nearside view in Figure 2, and also in Figure 3, you should be able to make out the 1,146 km diameter Mare Imbrium. This has an arc of mountains along its south-eastern edge, which is a relic of the basin rim. If you look at the west view in Figure 2, you will see the Orientale basin to the south of the equator. This has an inner ring about 500 km across partly filled by mare basalts within a 920 km outer ring that is mostly basalt free. The nearside maria are probably all double-ring, or multiple-ring structures like this, but have been so fully flooded that no trace of the inner rings is visible at the surface.

The Orientale basin straddles the boundary between nearside and farside. The largest mare entirely within the farside is Mare Moscoviense, which is in the upper left of the farside view in Figure 2. This fills the inner ring of a double-ring basin nearly 500 km in diameter, and spills out into part of the outer ring. There is a much larger basin on the farside that has little or no mare basalt within it. This is the 2,500 km diameter South Pole–Aitken basin, which is the largest and oldest basin preserved on the Moon. It has excavated to a depth of 13 km and shows up as an area of medium albedo (darker than highlands, but lighter than mare) stretching north from the south pole in the Figure 2 farside view.

Craters and dating

The Moon's craters have turned out to be very useful for working out the sequence of events that has affected the Moon. Working this out just requires common sense. For example, a crater that is superimposed on another must be younger, and (more usefully) the mutual relationship between dispersed ejecta from one crater and another that lies amidst the ejecta shows which of the two is youngest. There are also diagnostic relationships between craters and mare basalts, as Figure 5 illustrates.

There is a lot to see in this image. For now I will discuss just the relationships between the craters and the mare basalts that fill the whole of this field of view. The crater Beketov and those labelled D, E, and L are simple bowl-shaped craters, and there can be no doubt that these are younger than the mare basalt surface. However, what about Jansen? This is 24 km in diameter, easily big enough to have a central peak. However, it has a flat floor only about 150 metres below its rim. The explanation is that the Jansen crater was already there when the mare basalts flooded this area. The basalt lava flows overtopped the rim of the crater, flooding and burying its central peak. Subsequent degassing and thermal contraction caused the lava surface to subside, which allowed the rim of the buried crater (but not its more deeply buried central peak) to re-express itself on the topography. Crater Y punches 550 metres into Jansen's lava-flooded floor, and was made by an impact that happened much later.

Crater R, slightly larger than Jansen, is scarcely discernible at all. You can make out its circular outline within an otherwise normal lava surface. This is what is known as a 'ghost crater', denoting an old crater so completely flooded that it hardly shows up at all.

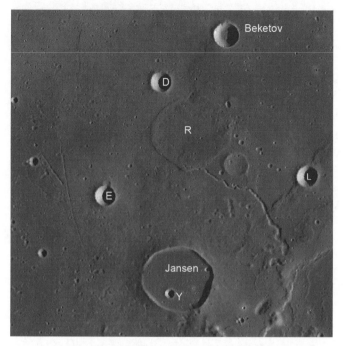

5. **A 110 km wide region in the north of Mare Tranquilitatis, including Jansen crater. The letters are the IAU designations of Jansen's 'satellite craters'.**

There is a smaller (unlettered) ghost crater just outside it, to the south-east.

The sequence of events that we can deduce from these simple observations is as follows. First, there was an impact that formed the large basin that is now occupied by Mare Tranquilitatis. Next, there were impacts into the floor of that basin, to make craters such as Jansen itself, Jansen R, and Jansen R's unnamed smaller neighbour. Then came the flooding of the basin by mare basalt lava flows, which presumably completely hid some craters but was not sufficient to fully obscure the craters now visible as 'ghosts'.

The new surface was subsequently struck by the impactors that made Beketov, Jansen D, E, L, and Y, and numerous smaller craters.

The lunar cratering timescale works on the basis that the longer a surface has been around, the more craters will have formed on it. This can be tested using mutual age relationships such as we have just looked at. It has been calibrated into an absolute timescale using samples of rocks and minerals brought back from the Moon that have been dated in the laboratory by measuring the accumulated products of radioactive decay. This shows that the rate of crater formation was intense during the interval 4.1–3.8 billion years ago, an epoch known as the late heavy bombardment. No lunar surfaces have survived from before this time, so we are not sure what was happening even earlier. The term 'late' denotes late in terms of the formation of the Solar System. In fact it was early in lunar history, the Moon and the Earth being about 4.5 billion years old.

Thirty known basins formed during the late heavy bombardment. The South Pole–Aitken basin is the oldest, and may have been made about 4.1–4.0 billion years ago. The Imbrium basin was formed about 3.8 billion years ago, and the Orientale basin (the youngest of its kind) is about 3.7 billion years old.

In all cases, it was several hundred million years after basin formation that the most extensive mare basalts were erupted. Most had been emplaced by about three billion years ago, but eruptions continued in some areas until about one billion years ago. Over the past 3.5 billion years the rate of cratering seems to have been fairly constant and much lower than during the late heavy bombardment, though it is impossible to rule out short flurries of bombardment.

Cratering has not ceased. As already mentioned, material in one of Tycho's rays (and hence Tycho itself) has been dated at a hundred million years old. Smaller, brand new craters have been imaged by lunar orbiters, including an eighteen-metre crater whose formation was marked by a brief flash recorded by ground-based telescopes on 18 March 2013. The impactor that made this crater would have been about a metre across, and would not have survived to make a crater if it had encountered the Earth, because it would have been vaporized by friction in our atmosphere.

Names of craters and maria

The naming of craters on the Moon is now under the control of the IAU. Most are named after deceased scientists and polar explorers. A few more recently named craters honour cosmonauts and astronauts who lost their lives. Names of maria are mostly Latin terms describing weather conditions (for example, Mare Imbrium means 'Sea of Rains'). One exception is Mare Moscoviense (Moscow Sea) that was named when discovered on the first farside images from Luna 3. Another is Mare Orientale (Eastern Sea), a name that might seem perverse for a feature on the central longitude of the Moon's *western* hemisphere. Part of it can be seen from Earth during favourable libration, and the name is inherited from a time when Earth-bound astronomers used directions in the Earth's sky rather than thinking about coordinates from the perspective of someone on the Moon's surface.

Given the history of the naming of Jupiter's moons, you will probably not be surprised to learn that early observers put forward conflicting schemes for lunar nomenclature. The present-day scheme derives from that of an Italian Jesuit named Giovanni Riccioli (1598–1671), published in 1651. The convention of applying letters to identify smaller craters within or around a larger named crater (such as Copernicius B in Figure 4 and Jansen D, E, L, R, and Y in Figure 5) was devised by the German Johann Mädler (1794–1874). Note that these letters recognize proximity alone. Jansen D, E, L, and Y must have formed long after Jansen itself, so their origin is certainly not related to Jansen and probably not to each other.

Mare basalts and regolith

The mare basalts appear to have formed when parts of the Moon's mantle grew hot enough for 'partial melting' to begin, enabling some melt to be sweated out, leaving a solid residue in the mantle. The magma that reached the surface was similar to basalt in composition. This has lower viscosity than most molten rock, and was able to spread far across the surface before solidifying.

The early stages of eruption may have been violent, with molten lava flung hundreds of metres skywards before falling to the ground. The main fissures from which the lavas were erupted cannot be seen, because they were filled in and covered by later more gentle effusions, but some buried fissures filled by dense solidified lava were identified in gravity mapping by NASA's GRAIL mission in 2012 (Table 2). Some later lava flows carved channels into the surface of earlier flows, and there is a nice example in Figure 5 beginning about 30 km from the eastern rim of Jansen and winding towards Jansen R. West of Jansen E you may be able to make out a couple of straight, narrow fissures, which overlie dyke intrusions or are cracks that formed as the lava cooled and contracted.

Although the Moon is rightly described as a rocky body, there is actually very little solid bedrock exposed at the surface. Impacts of all sizes have distributed a layer of fragments (estimated to be about ten metres thick on the highlands and five metres thick on the mare surfaces, which are younger) that makes up the lunar soil, more properly called 'regolith'. Most of the regolith is made of fragments less than a tenth of a millimetre in size, though impacts

Table 2 Highlights of lunar exploration

Name	Country	Date	Achievements
Luna 1	USSR	4 Jan. 1959	Fly-by, no pictures
Luna 2	USSR	13 Sept. 1959	Impact on to Moon
Luna 3	USSR	6 Oct. 1959	Fly-by; first farside pictures
Ranger 7	USA	31 July 1964	Impactor
Luna 9	USSR	3–6 Feb. 1966	Lander; first pictures from the surface
Luna 10	USSR	3 Apr.–30 May 1966	First lunar orbiter
Surveyor 1	USA	2 June 1966–7 Jan. 1967	Lander
Lunar Orbiter 1	USA	14 June 1966–29 Oct. 1967	Orbiter
Apollo 8	USA	24–7 Dec. 1968	First manned orbiter
Apollo 11	USA	20–1 July 1969	First manned landing; 21.5 kg of samples
Apollo 12, 14–17	USA	Nov. 1969–Dec. 1972	Manned landings; 360 kg of samples
Luna 16	USSR	20–4 Sept. 1970	First robotic sample return; 0.1 kg

(continued)

The Moon

37

Table 2 Continued

Name	Country	Date	Achievements
Lunokhod 1	USSR	17 Nov. 1970–14 Sept. 1971	First lunar rover; 11.5 km traverse
Luna 20, 24	USSR	Feb. 1972, Aug. 1976	Robotic sample returns 175 kg
Lunokhod 2	USSR	15 Jan.–11 May 1973	Lunar rover; 40 km traverse
Hiten	Japan	Mar. 1990–Apr. 1993	Orbiter/impactor
Clementine	USA	Feb.–June 1994	Orbiter
Lunar Prospector	USA	Jan. 1998–July 1999	Orbiter/impactor
SMART-1	Europe	Nov. 2004–Sept. 2005	Orbiter
SELENE (Kaguya)	Japan	Oct. 2007–June 2009	Orbiter/impactor
Chang'e 1	China	Nov. 2007–Mar. 2009	Orbiter/impactor
Chandrayaan-1	India	Nov. 2008–Aug. 2009	Orbiter and impactor
Lunar Reconnaissance Orbiter	USA	June 2009–	Orbiter
LCROSS	USA	9 Oct. 2009	Impactor
GRAIL	USA	Jan.–Dec. 2012	Gravity mapping from orbit
Chang'e 3	China	14 Dec. 2013–	Orbiter, lander, and rover (Yutu)

6. Apollo 15 astronaut James Irwin and the Lunar Roving Vehicle on the brink of Hadley Rille (a 1.5 km wide, 300-metre-deep lava channel) in July 1971. Note the footprints in the regolith in the foreground.

big enough to penetrate into bedrock fling out some boulder-sized pieces too, like the example in the left of Figure 6.

The dominantly dusty nature of the regolith means that footprints are sharp, and indeed the paths trodden by Apollo astronauts in the years 1969–72 can be seen today in super-high-resolution images taken from lunar orbit.

Lunar exploration

The Moon was the prime target for exploration until the US won the 'space race' with its landing of Apollo 11 in July 1969. After the ending of the Apollo programme and the Soviet unmanned sample return efforts (three landers that brought back just short of a third of a kilogram of regolith between them) in the 1970s, there was an interval of nearly twenty years before exploration picked up again, with a new generation of unmanned orbiters equipped for specific scientific observations. More nations joined in too, as can be seen in Table 2. Looking to the future, India plans an orbiter/lander/rover package named Chandrayaan-2 for 2016–17 and China plans a sample return mission named Chang'e 5 in 2017. An ambitious crowd-funded project called Lunar Mission One was begun in the UK in 2014, intended to land a probe to drill into the floor of the South Pole–Aitken basin in 2024. It is not clear when humans will return to the Moon, but if I had to place a bet it would be on a Chinese mission in about 2020.

'And don't forget to bring back some rock!' was the gag in a newspaper cartoon depicting someone waving goodbye to the Apollo 11 crew before launch. Bringing samples of the Moon back to Earth, where they can be subjected to detailed and precise analysis, was the most important scientific goal of the project. I have already mentioned radioactive dating. Other analyses include determining what minerals make up lunar rock and whether the abundances of different elements in those minerals differs from what we would find in rocks from Earth. For elements such as oxygen that have more than one stable isotope, the relative abundances of these isotopes can be used as a 'fingerprint' to see whether the Earth and Moon were made from the same source material. Even something as basic as examining a rock with a microscope to study its texture can tell us things about the rock's origin and history that can't be deduced from orbit.

The six Apollo landings collected a total of 382 kg of samples, and the three successful robotic Luna missions brought back a further 0.32 kg. All of these were from the nearside, so some important regions remain unsampled. The South Pole–Aitken basin is perhaps the most compelling target for future sample return.

However, there is more Moon rock available on Earth than what has been deliberately brought back. Impacts on to the Moon can throw a fraction of their ejecta out with sufficient velocity that it escapes the Moon and may eventually land on the Earth as a meteorite. Lunar meteorites were first recognized in 1982. They are different to the more common types of meteorites that come from asteroids, though if we didn't have previously known samples of the Moon to compare them with their origin might still be in doubt. About 48 kg of lunar rock has now been collected as meteorites. About half of this is likely to come from the farside, but of course we cannot identify the specific location where any example originated.

Some lunar rocks are breccias, made of pieces of basalt and/or highland rock smashed apart by impacts but also welded together by the heat generated by impacts. Other samples are lumps of a single basalt or highland rock type. Regolith contains fragmented or pulverized relics of all these, and also glassy beads about 0.1 mm in size that appear to be frozen droplets of basaltic spray from explosive eruptions.

Most mare basalts in and around Mare Imbrium show geochemical fingerprints of having been derived from a source region relatively rich in potassium (K), rare earth elements (REEs), and phosphorous (P), known as KREEP. It is thought that this records an anomalous patch in the nearside mantle in which heat from radioactive decay of potassium (and also thorium, which has been mapped from orbit by detecting its emitted gamma rays) was a key cause of the partial melting that led to the eruption of the mare basalts.

An early recognized characteristic of lunar samples shown by the analytical techniques available at the time was a lack of water bound up inside minerals, nor any chemical alteration of minerals. Mineral crystals billions of years old look as fresh as if they grew only last year, whereas on the Earth the damp environment would have led to chemical alteration penetrating into minerals along fractures and cleavage planes.

The Moon is a very dry place. In addition to scarcity of water, it has a much lower inventory of 'volatile elements', such as sodium.

On the other hand, the relative abundances of the three stable isotopes of oxygen in lunar rocks is an almost perfect match to what we find in the Earth, which suggests that the two bodies formed from the same source material. However, if that were the case, why does the Moon not have an iron core to match the Earth's core? Small core and depletion in volatiles on the one hand, and similar oxygen isotopes on the other, are apparent contradictions that have to be reconciled when trying to explain how the Moon formed in the first place.

The Moon's origin

In 1879 George Darwin (1845–1912), second son of his more famous father Charles, proposed that the Moon formed by fission—splitting off from a previously more rapidly spinning Earth. This could explain the Moon's lack of a core and oxygen isotope match to the Earth, but not its depletion in volatiles. A co-accretion model, in which the Earth and Moon grew side by side while the Solar System was forming, falls down on the lack of a core as well as mismatched volatiles. Capture of the Moon by the Earth after independent formation would be dynamically very difficult, and would require the oxygen isotope match to be a fluke.

The Moon's origin is still a matter of debate. A theory that it was formed by a 'giant impact' on to the early Earth was developed in the mid-1980s and has become widely accepted, because it could explain the matches and mismatches between the two bodies. The latter stages in the growth of terrestrial planets are probably a series of a few collisions (giant impacts) between bodies of roughly similar size (known as planetary embryos), rather than growth of a larger body by numerous impacts of much smaller bodies. Giant impacts usually result in a merger of the two planetary embryos, leaving a larger embryo expected to be largely molten because of the heat generated by the impact. This melting would make it easy for iron (which is dense) to sink inwards to form a core; a process described as differentiation.

According to the giant impact hypothesis of the Moon's origin, the final giant impact experienced by the proto-Earth was when a differentiated Mars-sized planetary embryo struck it a glancing blow. Rather than leading to full merger, the collision ejected the impactor's mantle and some of the Earth's mantle into space, whereas the impactor's core ploughed inwards and merged with the Earth's core. The Moon then grew in orbit round the Earth, from a mixture of the impactor's mantle and the target's mantle.

Energy converted to heat as the Moon grew would have melted it, allowing what little free iron there was to sink to form a tiny core. More importantly, as the Moon's global magma ocean began to cool, the first crystals to form would have been anorthite, the very mineral discovered to make up most of the lunar highlands. Anorthite has a relatively low density, and the crystals would tend to rise. Probably they would need to clump together to form larger masses before their buoyancy could overcome the magma's viscosity, but then they would rise until they breeched the surface, coalescing to form the lunar highland crust.

Radioactive dating has determined that the oldest lunar crust samples were formed about 4.35 billion years ago, about 200 million years after the birth of the Solar System, but the Moon-forming impact could have occurred as long ago as 4.5 billion years.

Water and individual volatile elements would have been preferentially lost to space before the debris from the giant impact was able to coalesce into the Moon, so that the giant impact hypothesis is the most credible story. The giant impactor has even been given a name—Theia, after the mythological Selene's mother. A recent variant of the hypothesis, invoked to explain some of the differences between nearside and farside, is that the debris originally formed two moons, which merged in a relatively slow mutual collision a few tens of millions of years after the giant impact.

Water on the Moon

The Moon is not so entirely bone dry as it seemed to be, based on the original interpretation of the Apollo data. This century, traces of water have been found within lunar samples in a type of mineral called apatite. Apatite's crystalline structure causes it to incorporate any available water molecules, and also any large negative ions such as those formed by fluorine and chlorine. There is an ongoing debate about how to use apatite to interpret the water content of the Moon's mantle. It could be anything from ten parts per million to one part per thousand. We aren't sure what the average water content of the Earth's mantle is either, but it is almost certainly higher than the Moon's, perhaps about 1 per cent, and so the Moon's surface rock and its interior are still regarded as much drier than the Earth.

However, there is a second, unrelated, reservoir of water on the Moon. This occurs as ice inside the cold, −170°C, craters near the poles whose floors are never illuminated by the Sun. Evidence for this built up slowly. In 1994 the Clementine orbiter showed that radio waves bounce from those craters in a way consistent with ice. Five years later, the neutron spectrometer on the Lunar Prospector orbiter demonstrated concentrations of hydrogen that could most reasonably be interpreted as residing in water (H_2O). The clincher came in 2009, when the Centaur rocket that had delivered the Lunar Reconnaissance Orbiter was crashed into the permanently shadowed floor of a crater named Cabeus near the south pole. A probe called LCROSS (Lunar Crater Observation and Sensing Satellite) followed six minutes behind. Before it too crashed, it was able to analyse the ejecta flung up by the Centaur crash, showing the floor of the Cabeus to consist of about 6 per cent water by mass.

There is a fairly simple explanation for the Moon's polar ice, and it has nothing to do with any water from *inside* the Moon that has been there since the Moon's birth. When a comet hits the Moon and makes a crater the comet's ice is vaporized. The water molecules are added to the Moon's exosphere. If a molecule hits a hot surface, it will bounce, and will eventually be lost to space. However, if it hits a cold surface it will stick there. If it hits a permanently cold place, it will stick there indefinitely. In this way permanently shadowed crater floors act as 'cold traps' where ice accumulates molecule by molecule, and continues to do so today. The same process builds up ice inside Mercury's polar craters too.

Chapter 3
The Moon's influence on us

The Moon's presence in the sky was a spur to our initial move into space, but it has long pervaded human culture in many other ways, including providing themes for music and song (both good and bad) and influencing language. For example, 'waxing and waning' refers to the Moon's illuminated face growing or shrinking as it goes through its phases.

The Moon has had almost as much influence as the Sun in the way we keep time. It takes the Moon 27.3 days to complete an orbit round the Earth, but because the Earth moves about a twelfth of the way round the Sun in that time, it takes 29.5 days for the Sun–Earth–Moon relationship to repeat. This 29.5-day cycle of phases is the origin of the months that have been used in virtually every known human calendar. In the English and Germanic languages the words for Moon and month even share the same root. Dividing a lunar month into four (for example, the time between first quarter and full Moon) is probably the origin of our seven-day week.

Western cultures, and for official purposes the whole world, use a solar calendar. This defines the start of the year by a fixed position in the Earth's orbit round the Sun. To keep things neat there are twelve months in a 365- or 366-day year, of which all except February are slightly longer than a lunar month. However, the

Islamic calendar is lunar, and counts time in years of 354 or 355 days (twelve 29.5-day lunar months), which are thus shorter than the Earth's orbital period round the Sun. Traditional Chinese and related oriental calendars are lunisolar, using lunar months but defining the new year as the second (sometimes the third) new Moon after the winter solstice, which means that years are of different length but average out to be equal to the Earth's orbital period.

The saying 'once in a blue Moon' originally referred to an infrequent or even impossible event. Recently, calendricists have formalized the expression 'a blue Moon' to refer to the occasion when a second full Moon occurs within the same calendar month, which averages out to about seven times in every nineteen years.

Although everyone should be used to seeing the Moon in the sky, there is a surprisingly common mistake. Visualize a crescent Moon, or better still ask an unsuspecting friend or a child to draw one. More than likely they will come up with a crescent resembling a letter ☾, rather than its mirror image ☽. They will do this even if they live north of the tropics, where the Moon looks like a ☾ when it is a waning crescent, which can be seen only in the dawn sky. I have no idea why this is. Most of us see the crescent Moon far more often in the evening sky, when from the northern hemisphere its shape is a ☽.

Seen from the southern hemisphere, objects in the sky look 'upside down'. There, the waxing Moon in the evening sky looks like a ☾, whereas the waning Moon in the morning sky is the other way round. The ☾ and ☽ shapes that I have described are angled, with the illuminated arc tilted downwards (towards the Sun, near or below the horizon), and the tilt becomes greater the closer you are to the tropics. From a viewpoint near the equator, if you see a crescent Moon in a dark sky it will always look like a ◡ (never ◠), because the illuminated edge has to be facing towards the Sun, which is below the horizon, either having recently set or being about to rise.

Ocean tides

Tides in the oceans are caused by the gravitational pull of the Moon and the Sun on the ocean water, which can respond much more freely than the solid Earth. The Moon is far less massive than the Sun, but this is more than compensated by its relative closeness to the Earth, with the result that the tidal force exerted by the Moon is just over twice as strong as the solar tide.

When the Moon and Sun are lined up at new Moon there are two high tides and two low tides in the day. These will occur close to noon and midnight, except where coastlines complicate the movement of water (for example round the British Isles). An important point is that there are two tides per day, not one, because the solid Earth is pulled towards the Moon and Sun while being pulled away from the ocean water on the nightside. The tidal effect is exactly the same at full Moon, because although the Moon and Sun are pulling in opposite directions it is the *difference* in the strength of their pulls on opposite faces of the Earth that determines the tide. At such times, because the solar and lunar tides add together, the tidal range (the amount by which the tide rises and falls) is greatest. This situation is referred to as a 'spring tide', a term that refers to rapid rise and is nothing to do with the season. Conversely, when the Sun and Moon are at right angles to each other in the sky (at first quarter and last quarter), the weaker solar tide acting six hours out of phase with the lunar tide decreases the tidal range. This situation is called a 'neap tide'.

These effects are summarized in diagrammatic form in Figure 7. The solid bodies of the Earth and Moon are distorted by the same forces, but by a much smaller amount than the oceans.

7. Sun, Earth, and Moon (not to scale) at four successive times to illustrate ocean tides. The ellipses round the Earth show the strength of the tide-producing force from the Moon (the larger ellipse) and from the Sun (the smaller ellipse). The effects add at new and full Moon to produce the highest tidal range (spring tides), whereas at first and last quarter the maximum solar and lunar tides are different by 90°, resulting in a smaller tidal range (neap tides).

If you are familiar with tides from seaside holidays, you may be surprised to learn that the tidal range in the deep oceans is only a few tens of centimetres. Coastlines can amplify the tidal range considerably. For example, along much of Britain's east coast the average tidal range is about four metres, but it can reach three times as great in Atlantic-facing 'funnels' such as the Bristol Channel and the Gulf of St Malo. Because the Moon is always progressing round the sky, the time between successive high tides is about twelve hours twenty-five minutes, so if you are on a beach waiting for the tide to turn it will happen later each day.

The elliptical shape of the Moon's orbit means that the strength of the lunar tide is not constant, but is greatest when the Moon is closest to the Earth. The strongest spring tides therefore occur when perigee coincides with either new Moon or full Moon *and* when the Moon is on the part of its orbit that intersects the plane of the Earth's orbit round the Sun, so that the solar and lunar tidal forces act in *exactly* the same direction. These effects are of only minor importance on most coastlines, because they can be outweighed by other effects such as an onshore wind pushing water coastwards, or low atmospheric pressure allowing the sea to rise more than normal.

Eclipses

The elliptical nature of the Moon's orbit affects eclipses too. When a new Moon coincides with the Moon crossing the plane of the Earth's orbit, the Moon will be exactly in line with the Sun as seen from somewhere on the globe, causing a solar eclipse. It is a lucky coincidence that (for most of the time) the Moon appears big enough to blot out the whole solar disc for up to seven minutes, allowing the Sun's otherwise invisible outer atmosphere (the corona) to be seen in all its glory. However, when a solar eclipse happens with the Moon at apogee, the Moon is too far away to completely hide the Sun, so at the time of exact alignment an annular eclipse is seen, with a bright ring of the Sun's disc surrounding the part obscured by the unlit Moon.

Because the Moon is relatively close to the Earth, parallax means that you have to be in the right place to see a total eclipse. The path of the Moon's shadow across the globe is only about 250 km wide even at perigee. Outside of this path, you can see only a partial eclipse where the Moon hides only part of the Sun.

On the other hand, when the Moon is exactly opposite the Sun in the sky, the Earth casts a shadow big enough to cover the whole Moon. This is called a lunar eclipse, and can be seen from anywhere where the Moon is above the horizon. Although the eclipsed Moon is in the Earth's shadow, it remains visible because it is lit by reddened sunlight that has been bent through the Earth's atmosphere. Looking at the Moon from the Earth, you see a dull red disc. Looking from the Moon, you would see a red 'sunset glow' rimming the Earth.

Orbital recession and day length

Just as the Earth raises tidal bulges on the Moon, so the Moon's pull raises tidal bulges in the solid shape of the Earth, in addition to the ocean tides. Because the Moon is so much less massive, its effect on the Earth is much less than the Earth's effect on the Moon. Thus the Moon has not managed to slow the Earth's rotation to match the Moon's orbital period (Pluto and its largest moon Charon are a rare example where this has happened).

Measurements show that the Earth's rotation is slowing by about 1.6 milliseconds per century, which is mostly a result of tidal drag from the Moon. The same forces are slowing the Moon's orbital speed (and, at the same time, its rotation). Slower orbital speed requires it to be further from the Earth, and it is receding at a rate of 3.8 cm per year. Recession of this sort can be calculated, but it has also been demonstrated by determining the gradually lengthening travel time of light between the Earth and Moon by directing laser beams on to retroreflectors left on the lunar surface by Apollos 11, 14, and 15 and the Soviet Lunokhods.

If the Moon's present-day rate of recession had been constant, the Moon would have been too close to the Earth to be stable between one and two billion years ago, so it seems that recession must be faster now than in the past. We have no way of measuring how close the Moon was in the distant past, but there are indications of how fast the Earth was spinning. Some varieties of coral show daily growth lines when examined under a microscope. In corals that are 370 million years old, these lines come in annual cycles of 400, showing that there were 400 days per year. The Earth's orbital period is unlikely to have changed by much, so we can conclude that the day length was shorter (about twenty-two hours) in order to fit 400 days into a year.

The Moon's influence on human behaviour

The very word 'lunacy' is derived from the Latin *lunaticus*, meaning a person affected by a supposed Moon-induced madness. However, whenever a sufficiently careful test has been done there has been found to be no correlation between the phase of the Moon and mental illness, or indeed any kind of human behaviour including crime, suicide, or birth rate. There is no physical reason to expect any influence. For example, the tidal forces experienced by the human body are vanishingly small.

It has been widely believed in most cultures that the Moon has an influence on human fertility, probably because of the similar length of the lunar month and the human menstrual cycle. However, this seems to be a coincidence, and no connection between the two has been shown to exist.

There is, however, a modern example of the Moon influencing human behaviour, and this is over the matter of so-called 'supermoons'. The Moon's angular size in the sky is about 14 per cent greater at perigee than at apogee, so that when perigee coincides with full Moon, the full Moon is bigger than average. The difference is too slight to notice unless you actually take some measurements, but despite this, over the past few years whenever full Moon falls close to perigee there has been a fuss on social media websites and in the conventional news media too, claiming that the phenomenon will be spectacular or even that it may cause a natural disaster of some sort.

The term 'supermoon' was coined in the 1980s by an astrologer, to denote full Moon occurring when the Moon is in the 10 per cent of its orbit that is closest to perigee. This catchy term has caught the public imagination and has replaced the more cumbersome phrase 'perigee full Moon'. However, it is misleading. A supermoon is only slightly bigger than average, and it is not even a particularly rare phenomenon (it is much more common than a 'blue Moon'). About one full Moon in every ten qualifies as a supermoon, and in fact there were three in succession in July, August, and September 2014. Contrast the commonness and trivial difference in apparent size of a supermoon with the amazing powers possessed by the fictional 'Superman', or with a 'supervolcano eruption'. The latter denotes a catastrophic eruption that produces at least ten times more ash than any volcano has managed in the past 1,000 years, and is an event that happens somewhere on Earth only once in about every 50,000 years on average. No correlation has been found, or is likely, between supermoons and events such as eruptions and earthquakes.

Although its angular size in the sky is bigger at perigee, the Moon's surface isn't actually any brighter then. Surface brightness depends on the Moon's distance from the Sun, not its distance from the Earth. The dominant factor is the eccentricity of the Earth's orbit, which leads to a roughly five million-kilometre difference in Earth–Sun (and Moon–Sun) distance between nearest (in early January) and furthest (in early July). This swamps the 40,000 km variation in Earth–Moon distance, and leads to a 7 per cent range in the Moon's surface brightness. However, because the Moon's angular size in the sky is greater when it is closest, for a given Earth–Sun distance full Moon at perigee does provide about 30 per cent more moonlight than when a full Moon occurs at apogee.

The human eye adapts to different levels of brightness. Our pupils dilate or contract to allow a comfortable amount of light on to our retinas, but our memories do not store this information. Thus, without using instruments we have no way to make a quantitative comparison between one full Moon and another several weeks or months later, and anyone who just looks at a supermoon and claims that they can tell that it is bigger or that the moonlight is brighter than normal is fooling themselves.

In contrast to the hyped-up and almost entirely imagined supermoon phenomenon, there is an unrelated optical illusion that can genuinely make the Moon look big. Whenever we see the Moon low in the sky (the closer to the horizon the better), the Moon seems to be large. It isn't really any bigger, and if you measure it the size doesn't change. This is an effect called 'the Moon illusion', and seems to be a result of the way your brain processes information. When the Moon is high in the sky, it seems small—isolated and adrift in a sea of back. However, when it is near the horizon there are distant but familiar objects with which to relate it, such as trees and rooftops, beyond which the Moon looms large.

The Moon and life on Earth

The Moon affects the behaviour of wildlife in numerous ways. Nocturnal land animals tend to vary their activity with the phase of the Moon, according to whether the extra light around full Moon is helpful or harmful to their success as predators or safety as prey. Many marine species use the Moon as a clock to trigger mass spawning, not because the light level matters, but because the success of this reproductive strategy depends on synchronization. Some turtles wait for spring tides so they can come ashore easily and lay their eggs somewhere that will stay dry until their young hatch.

The Moon may have influenced life on Earth in more fundamental ways than merely affecting feeding and breeding habits. It has been suggested that without the Moon to cause tides, it would have been much harder for life to migrate from the sea on to land. That's probably wrong, because the Sun on its own would cause twice-daily tides with about half the range that occurs in the average combined lunar and solar tides. This would offer plenty of scope for marine organisms to find themselves temporarily stranded by a falling tide.

Another influence of the Moon is that its presence might stabilize the tilt of the Earth's axis. Currently this is 23.4°. The axis points to a direction in space that changes only very slowly, and this tilt is responsible for the seasons as the Earth progresses round its orbit. Calculations suggest that over the past five million years the tilt has varied by about 2.5° with a period of about 41,000 years. This affects the climate, as can be seen in the fossil record. Studies in the 1990s suggested that if the Earth had no Moon, our axial tilt would experience much wilder fluctuations, ranging from zero to nearly 90°. This would have led to major extremes of climate, even worse than those experienced by Mars, which has only tiny moons and where axial tilt is currently varying between about 5° and 45°.

Such extremes of climate could have made it impossible for advanced life to establish itself on land, in which case if the Moon did not exist then we wouldn't either. However, more recent studies have contested the significance of the Moon, and suggested that the Earth's axial tilt could stay within narrow bounds even without the Moon's influence.

Moonbases and lunar resources

When humans eventually return to the Moon, it will probably lead to a more sustained lunar presence than the few days achieved by each Apollo landing. There are many good scientific reasons for sending trained astronauts (preferably geologists) if the goal is to explore the Moon. Humans can use their time much more efficiently than remote-controlled rovers. In 1973 it took the Soviet Lunokhod 2 nearly four months to travel further than the Apollo 17 astronauts managed in three excursions totalling twenty-two hours.

Apart from learning more about the Moon, being on the Moon would enable other science too. On the farside, the Moon shields you from radio interference from the Earth, making a farside radio telescope an attractive prospect for astronomers. Bizarrely, the Moon may also be the best place to find evidence of conditions on the early Earth. Just as impacts on to the Moon fling off ejecta that can be collected as meteorites on the Earth, so impacts on to the Earth must throw out ejecta that could be collected on the Moon. A piece of the Earth's surface that was ejected on to the Moon three billion years ago would be a wonderful find, because it would be in pristine condition compared to any weathered relict still on Earth. We would hope to find tiny but analysable samples of the early Earth's atmosphere trapped as bubbles within shocked glass, and maybe microfossils documenting the first stages of life on Earth.

It is debatable whether the Moon has any resources that it would make economic sense to bring back to Earth. The cost of getting to the Moon in order to fetch anything is very high. Because the Moon is so dry, it would be surprising to find ore minerals concentrated by the action of aqueous fluids circulating through the crust like we do on Earth, but there might be tempting concentrations of platinum-group elements hidden within craters made by the impact of metallic asteroids.

One commodity that it might make sense to export from the Moon to the Earth is an isotope of helium called helium-3. This is rare on Earth, but samples from the Moon show about ten parts per billion of helium-3 in lunar regolith, believed to have been implanted there by the solar wind. Helium-3 and a heavy isotope of hydrogen ('deuterium', which can be extracted from seawater) are the fuel required for nuclear fusion reactors, a proposed 'clean' source of electric power. The technological viability of fusion power has yet to be demonstrated, so all this is rather speculative. However, if a market for helium-3 ever does emerge on Earth, it would become commercially viable to strip-mine the surface regolith of the Moon, heating it to drive off, and then collect, the helium-3 gas.

Using the resources of the Moon *on* the Moon is a different matter. Water extracted from a permanently shadowed crater could make a nearby Moonbase much cheaper to run. Solar panels on a hilltop beside a polar crater could gather uninterrupted solar power that could be used to heat the regolith and to form it into building blocks, or to drive more complex processes to extract oxygen and metals, or to make glass fibre. Even in the case of habitat units brought from Earth, it could make sense to simply pile regolith over them to offer protection from solar storms.

Who owns the Moon?

There is no agreed legal framework to establish ownership of the Moon or any part of it. A United Nations 'Agreement Governing the Activities of States on the Moon and Other Celestial Bodies', instigated in 1979 and known for short as the Moon Treaty, declares that the Moon should be used for the benefit of all states and all peoples of the international community. It also expresses a desire to prevent the Moon from becoming a source of international conflict. These are fine sentiments, but the Treaty is toothless. Of the nations capable of flying independently to the Moon, only India has signed (though not ratified) the Treaty.

The Treaty also forbids ownership of extraterrestrial property by any organization or person, unless that organization is international and governmental. If this ever became effective it would be a major hindrance to the commercially driven exploration of the Moon and the rest of the Solar System. A more pragmatic approach might be to accept that if someone invests in exploration and can find a market for a resource, then they should be allowed to sell it to whomever wants to buy it—whether on-planet or off.

A claim, of sorts, to small areas of the Moon for non-commercial motives is enshrined in the Apollo Lunar Legacy Bill, which went before a US Congressional Committee in 2013. This seeks to give protected status to the wheeltracks, footprints, and hardware at the six Apollo landing sites. This seems reasonable to me. These are part of humanity's communal heritage, and it would be awful if some 'space pirate' could with impunity mess up the sites, or hack away bits of the hardware to sell privately on Earth.

8. Private property on the Moon? Top, the Lunar 21 lander, imaged from the Lunokhod 2 rover in January 1973. Bottom left, Luna 21 and right, Lunokhod 2: imaged in 2010 and 2011 by the Lunar Reconnaissance Orbiter.

It's a grey area, but most legislators would agree that nations or organizations that have landed an object on the Moon still own that object, though not necessarily the ground on which it stands. In 1993 the Lavochkin Association, a Russian aerospace company, sold the Luna 21 lander (Figure 8) and the Lunokhod 2 rover, both of which it had built, at a Sotheby's auction. These fetched a price of $68,500 and the purchaser was the British/American Richard Garriott (1961–), a video game developer and entrepreneur, who later made a self-funded visit to the International Space Station.

Chapter 4
The moons of giant planets

The giant planets Jupiter, Saturn, Uranus, and Neptune each have an extensive entourage of moons. Naturally, the first moons to be discovered were the largest, and are often called the 'regular satellites', but they are only part of the story. Not all the moons of giant planets can be neatly pigeonholed, but the overall situation is as follows.

Closest to the planet are small inner moonlets, mostly less than a few tens of kilometres in radius and irregular in shape. They are closely associated with the planet's ring system and their orbits are circular, lie in the planet's equatorial plane, and have radii less than about three times that of the planet itself.

Next are large regular satellites exceeding about 200 km in radius, which is large enough for their own gravity to have pulled them into near-spherical shapes, a condition described as 'hydrostatic equilibrium'. Their orbits are only slightly less circular than those of the inner moonlets, and have radii up to twenty or thirty times that of the planet. These too lie pretty close to the plane of the planet's equator.

Finally there are the irregular satellites, mostly less than a few tens of kilometres in radius. The term refers both to their irregularity in shape and to their orbits, which can be strongly elliptical and are usually considerably inclined relative to the planet's equator. They extend to about 400 times the radius of Jupiter and Saturn, over 800 times the radius of Uranus, and nearly 2,000 times the radius of Neptune.

Inner moonlets and all regular satellites except for Neptune's Triton travel round their orbits in the same direction that their planet rotates, which is described as 'prograde' motion. Most irregular satellites, as well as having inclined orbits, travel round their orbits in the direction opposite to their planet's spin. This is described as 'retrograde' motion, and has implications for these moons' origins.

Ice

Generally speaking, large moons are not rocky bodies. In the 1950s, telescopes fitted with spectrometers to measure the characteristics of reflected sunlight and trained on the larger moons of the giant planets began to reveal the presence of frozen water on most surfaces. This was not really surprising, because these bodies are a long way from the Sun, and mean surface temperatures are about −160°C for the moons of Jupiter, −180°C at Saturn, −200°C at Uranus, and −235°C at Neptune. At such extremely low temperatures, ice is mechanically very strong and behaves like rock; it can sustain craters, cliffs, and mountains without flowing like a glacier would on Earth.

Just as importantly, the temperature was also very low when these moons were forming. Jupiter, five times further from the Sun than is the Earth, lies beyond the 'ice line'. Temperatures here were low enough to allow water to condense directly into ice from the gas cloud surrounding the young Sun. Bodies that formed beyond the ice line generally contain more ice than rock, because in the cloud of gas and dust from which the Solar System formed the elements required to make water (hydrogen and oxygen) were more abundant than the key ingredients of rock (silicon and various metallic elements plus oxygen). Where hydrogen could form solid compounds, it did so, so rock dominates only inside the ice line.

Carbon and nitrogen are common elements too, and these went to make up other varieties of ice that condensed further from the Sun. Possibly at Saturn, and certainly at Uranus and beyond, the ice is not just water but is mixed with frozen ammonia (NH_3), methane (CH_4), carbon monoxide (CO), and (at Neptune) even frozen nitrogen (N_2). In the giant planets much of this, especially water, occurs as ice in their interiors below a thick gassy envelope mainly of hydrogen and helium, but the moons have too little gravity to have collected a lot of gas, so ices dominate.

The abundance of ice explains the low densities of most moons. The tables in the Appendix list densities of between 1,000 and 2,000 kg per cubic metre for most regular satellites of the giant planets. A rocky body should have a density of more than 3,000 kg per cubic metre, whereas water ice has a density of 1,000 kg per cubic metre (other ices are even less dense). Thus the lower its bulk density, the more ice and the less rock a moon contains.

Missions to moons

There would be much less to say were it not for space probes that have visited the giant planets and their moons. Exploration began with fly-bys (missions that flew past the planet) but has moved on to the stage of orbital tours in the case of Jupiter and Saturn, which have each had a mission that orbited the planet for several years and that was able to make repeated close fly-bys of at least the regular satellites. Close fly-bys of moons enable detailed imaging, and usually take the probe close enough to see how the moon affects the strong magnetic field surrounding the planet and to detect whether the moon also has its own magnetic field. The size of the slight deflection to a probe's trajectory as it passes close to a moon enables the moon's mass to be determined. Knowing the moon's size, it is then easy to work out its density.

The story begins with NASA's Pioneer 10 that flew past Jupiter in December 1973, and Pioneer 11 that flew past Jupiter in December 1974 and then Saturn in September 1979. These were concerned mostly with the planets' atmospheres and magnetic fields, and collected little data about their moons.

It was NASA's two Voyager probes that really opened our eyes to the moons. Voyager 1 flew through the Jupiter system in March 1979 and through the Saturn system in November 1980. In August 2012 it became the first space probe to cross the heliopause, where the solar wind fails, and to enter interstellar space. Voyager 2 made fly-bys of all four giant planets: Jupiter in July 1979, Saturn in August 1981, Uranus in January 1986, and Neptune in August 1989. It remains the only probe to have visited Uranus or Neptune.

NASA's Galileo mission went into orbit around Jupiter in December 1995. After dropping an entry probe into Jupiter it toured the moons for eight years until it ran out of manoeuvring propellant and was allowed to crash into the planet. There was a serious early set back because its main communications antenna, a parabolic dish, failed to deploy. This meant that data had to be transmitted using the backup 'low-gain' antenna, reducing the total number of images that could be collected, but in-flight programming and data compression techniques rescued much of the science.

The joint NASA–European Space Agency (ESA) mission Cassini–Huygens arrived at Saturn in June 2004. It released the Huygens lander that parachuted to the surface of Titan in January 2005, while the Cassini orbiter began a long and complex orbital tour that is scheduled to end with entry into Saturn's atmosphere in 2017.

Cassini flew past Jupiter in December 2000 on its way to Saturn, and for several days was able to complement the Galileo orbiter's observations of volcanic eruptions on Io. More images of these spectacular events were provided by NASA's New Horizons mission, which made a close pass by Jupiter in February 2007 on its way towards Pluto, which it flew past in July 2015.

Jupiter's regular satellites

Jupiter's four Galilean moons are the archetypal regular satellites. I consider them here as a family, reserving individual treatment for Chapter 5. They are shown together in Figure 9, cut away to reveal their internal structures. These were deduced mainly from clues to their internal density distribution achieved by Galileo fly-bys together with measurements by Voyager and Galileo of the interplay between each moon and Jupiter's magnetic field. The latter shows that Jupiter's magnetic field induces a field within Europa and Callisto, most likely achieved by electrical conduction in a salty internal ocean. Ganymede has a fairly strong magnetic field of its own, which may be generated by convection currents acting like a dynamo in a liquid iron sulfide outer zone of its core, as happens inside Mercury and the Earth. Io's magnetic field has been less well characterized, and we cannot be certain whether it results from motion in a fluid core or is an induced field with a relatively shallow source. Three are differentiated bodies, in which the denser material has been able to segregate inwards to form a core, but Callisto lacks a strong internal density gradient, showing that it is only weakly differentiated.

Io is the densest moon in the Solar System and is the only regular satellite to lack surface ice. It can be thought of as a larger (and more active) version of our own Moon. It has a rocky surface, stained yellow and red by sulfur compounds distributed by ongoing volcanic eruptions.

Europa is a smaller (and less active) version of Io, buried by water, which is solid near the surface (ice) and liquid at depth where it forms a global ocean. Europa is nearly as dense as the Moon, and

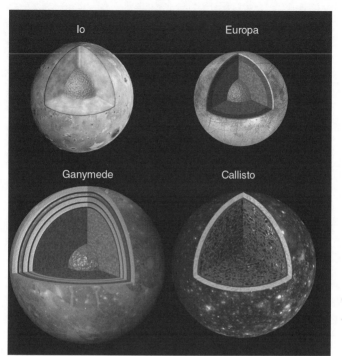

9. **Cutaway views to show the inferred internal structures of Jupiter's regular satellites, shown to scale. Io, Europa, and Ganymede are shown with an iron-rich core, surrounded by rock. The outer layers of Europa, Ganymede, and Callisto are ice, and the dark layers inside are liquid water. Callisto's interior is shown as an undifferentiated mixture of rock and ice.**

this shows that the shell composed of water (ice plus liquid) is only about 100 km thick. Europa's surface ice is able to fracture and migrate relative to the interior, but neither Europa nor any other moon shows behaviour closely similar to the formation and migration of tectonic plates as on Earth.

Ganymede is the most massive moon in the Solar System and is the only differentiated Galilean moon in which the internal pressure is sufficient to compact H_2O-ice into denser crystalline structures. Its surface ice is the normal sort, known as Ice I, but at greater depths the calculated pressure is enough to compress it into phases known as Ice III, then Ice V, and finally Ice VI. Figure 9 shows a complex internal structure proposed for Ganymede in 2014. According to this model, each ice phase has an underlying liquid layer, giving the appearance of a multi-layered sandwich. If these liquid layers exist, they are probably quite strong brines, containing salts dissolved out of the rocky interior that keep them liquid at temperatures too cold for pure water to melt.

The density of the Galilean moons decreases outwards from Jupiter. This is a strong clue to their origin, and suggests that they grew from a cloud of debris around the young Jupiter (in much the same way as the planets grew around the Sun), and that the heat radiated by Jupiter was sufficient to starve Io, and to partially starve Europa, of the water that was available to moons further out. Debris around Jupiter would have shared Jupiter's rotation, which would naturally result in these having moons prograde orbits close to the planet's equatorial plane, as indeed we see.

Orbital resonance and tidal heating

When the Voyager missions were being planned, the regular moons were expected to be fairly dull places. It was argued that, being largely icy and relatively small, the amount of heat-producing radioactive elements contained in any interior rock would be too little for there to have been any internally driven activity that could have affected their surfaces during the past three or four billion years. They would therefore be 'dead' worlds, heavily scarred by impact craters like the lunar highlands. Rock or ice, it doesn't matter: impacts cause craters that look much the same.

However, it turns out that Io has a surface so young that no impact craters at all have been seen there, and Europa has very few. There are plenty of impact craters on Ganymede and Callisto, some of which you should be able to make out on Figure 9, but Ganymede has tracts of paler, younger terrain cutting across its surface. So, as well as an outward trend of decreasing density, the Galilean moons have an outward trend of increasing surface age.

The nearer moons are not actually any younger, but they have been resurfaced more recently by geological activity. The explanation lies in their orbits. Tidal friction has long since slowed down their spin, so they have synchronous rotation matching their orbital periods. It has also made their orbits become much more circular than the Moon's, because they are orbiting a much more massive planet with stronger gravity.

In an elliptical orbit, libration would displace a moon's tidal bulges to and fro about their mean position. The distortion of the interior to allow this to happen must add heat by means of internal friction, and would encourage the interior to become differentiated, if the even more powerful tidal heating before its rotation became synchronous had not already done so. If you want to experience the efficacy of internal frictional heating, try bending a wire coat hanger to and fro, and then (carefully!) touch the bent part to your lip.

If Jupiter had only one moon, the planet's pull would have taken less than a hundred million years to force the moon's orbit into an exactly circular shape, whereupon there would be no more libration and no tidal heating. However, as you know Jupiter has four regular satellites. While Voyager 1 was speeding towards Jupiter, the American Stan Peale (1937–2015) and colleagues were computing the effect that membership of a family of moons can have on tidal heating.

The orbital periods of Europa and Ganymede are exactly twice and four times that of Io, so that for every four circuits of Jupiter made by Io, Europa completes exactly two and Ganymede exactly one. This is a situation described as 'orbital resonance', and has been brought about by mutual gravitational interactions between the moons. This means that Io passes Europa at exactly the same point in its orbit every time, and Europa passes Ganymede at exactly the same point in its orbit. The repeated slight gravitational tug between moons has prevented their orbits from becoming *exactly* circular. If you saw them drawn on paper they would look like circles, but there is enough departure from circularity for the orbital speed to vary so that the tidal bulges migrate slightly to and fro, and swell and contract in height, heating their interiors in doing so.

Peale and his colleagues calculated the amount of heating that this process should produce inside Io. Having made the best assumptions they could about Io's internal structure and strength, they published their findings in the 2 March 1979 issue of the journal *Science*, writing 'dissipation of tidal energy in Jupiter's satellite Io is likely to have melted a major fraction of the mass. Consequences of a largely molten interior may be evident in pictures of Io's surface returned by Voyager 1.' This is the most remarkable example of timing in planetary science known to me; images recorded by Voyager 1 six days later revealed erupting volcanoes on Io, and a total lack of surviving impact craters.

Orbital resonance is a complex situation. While in a state of resonance, the amount of forced eccentricity (and hence the rate of tidal heating) can wax and wane, and also moons can drift in an out of resonance over millions of years. Present-day tidal heating accounts for Europa's young surface, and past episodes of tidal heating can be invoked for Ganymede and several regular satellites of other giant planets.

Other regular satellites

The regular satellite families of the other giant planets lack the clear outward trend of decreasing density seen at Jupiter, and, with the exception of Saturn's Titan, the individual moons are smaller. However, the orbital characteristics at Saturn and Uranus are similar; almost circular orbits in their planet's equatorial plane, even at Uranus where some catastrophe long ago tipped the whole system on to its side (Uranus' rotation axis is tilted at 97.9° relative to its orbit round the Sun). At present there is orbital resonance between some of Saturn's moons but none among Uranus' moons, though some surfaces bear signs of tidal heating suggesting that resonances occurred in the past.

Saturn has seven regular satellites (see Appendix). Working outwards these are Mimas, Enceladus, Tethys, Dione, Rhea, Titan, and Iapetus. Mimas is in 2:1 resonance with Tethys, though this seems to have produced no effective tidal heating. Enceladus is in 2:1 resonance with Dione, and this must power the eruption plumes that Cassini discovered near Enceladus' south pole that are discussed in Chapter 5.

Uranus has five regular satellites: Miranda, Ariel, Umbriel, Titania, and Oberon. Of these, Ariel and Titania are crossed by large fractures best attributed to ancient tidal heating. Miranda has a complex surface history, possibly the result of a previous 3:1 orbital resonance with Umbriel. It is slightly non-spherical (its radii are 240, 234.2, and 232.9 km) and some might regard it as Uranus' outermost inner moonlet.

These all probably formed around their planets in a similar manner to that proposed for Jupiter's Galilean moons, but the same cannot be said for Neptune's largest moon, Triton (not to be confused with the similarly named Titan at Saturn). This has an inclined and retrograde orbit. There is no known way in which Triton could have formed alongside Neptune and ended up in such an orbit, so it was probably captured by Neptune from elsewhere.

Most likely Triton began as a member of the 'Kuiper belt', a family of icy bodies, from 1,500 km downwards in size, that orbit the

Sun near Neptune's orbit and beyond. If this is correct, Triton's originally independent orbit round the Sun must have, on one occasion, brought it close enough to Neptune to become captured. Capture is very difficult to achieve, because there is too much momentum to dispose of. Usually the approaching body would just swing past the planet, or (rarely) collide with it. However, if the incoming object is actually double (a primary object and a moon), one can be captured and the other can be flung away faster than it arrived, with the total momentum of the system being conserved.

Such double objects can in fact be seen today in the Kuiper belt. Pluto and its large moon Charon are a prime example. Triton is slightly bigger than Pluto, but we do not know whether or not it was the larger or smaller component of the supposed double object that arrived at Neptune. Whether double or single, Triton's arrival and capture would have scattered any pre-existing regular satellites, which must have been lost to the Kuiper belt or destroyed in mutual collisions.

Trojan moons

When a smaller body orbits a larger one, there are two stable points where an even smaller body can reside or about which it can oscillate. These occur 60° ahead and 60° behind the orbiting body, sharing the same orbit about the larger body. Technically these are called Lagrangian points, and are designated as L_4 (ahead) and L_5 (trailing). However, they are often referred to as the 'Trojan points' because there are groups of asteroids named after characters from the Trojan war that share Jupiter's orbit round the Sun, clustered around its L_4 and L_5 points.

Saturn has four small moons in leading and trailing Trojan relationships with two of its regular satellites, Tethys and Dione. Calypso, Telesto, and Helene, which are between 10 and 20 km in radius, were discovered telescopically in the 1980s and 1990s. The

10. The Solar System's four known Trojan moons, imaged by Cassini and shown at approximately correct relative scale. Helene has a mean radius of 17 km. Polydeuces has never been seen in sufficient close up to reveal surface details. Helene and Polydeuces are Trojan co-orbitals of Dione, whereas Telesto and Calypso are co-orbitals of Tethys.

smallest, Polydeuces, which is less than 3 km across, was discovered on images taken by Cassini in 2004.

Images from Cassini (Figure 10) have revealed some surprising aspects to the surfaces of these Trojan moons. They have rather few impact craters, and so may be relatively young (this could still mean they are more than a billion years old). The figure shows the side of Helene that faces away from Saturn and its rings; the Saturn-facing side has more craters. Calypso's surface is the most reflective in the entire Solar System, but this is not apparent in Figure 10 as the images have been processed to show each surface equally well. This may be a result of Calypso sweeping up ice crystals erupted from Enceladus. Both Calypso and Helene show curious gullies on their surfaces. There is no likely liquid that would be stable under their surface conditions. Some kind of dry avalanche process may be responsible, but it is a mystery how this would work on a body whose surface gravity is only 0.02 per cent as strong as the Earth's.

Saturn is the only planet known to have Trojan moons, but this may be a result of observational bias. The Galileo orbiter didn't search for them at Jupiter, which could have small ones like Polydeuces. Voyager 2 didn't have chance to look for any at Uranus and Neptune, where even Calypso-sized Trojans would be hard to spot telescopically from Earth.

Irregular satellites

There are more known irregular satellites than any other class of moons. Jupiter has fifty-nine. The inner seven are in prograde orbits up to 238 Jupiter radii (Jupiter's equatorial radius, 71,492 km) in size, and the others are all in retrograde orbits extending out to 400 Jupiter radii. This is a consequence of the differential long-term stability of orbits relative to the size of a planet's Hill sphere (the range out to which the planet's gravity outcompetes the Sun's gravity). Prograde orbits are stable over billions of years out to only about half the Hill sphere radius, whereas retrograde orbits can be stable out to about two-thirds of the Hill sphere radius. Jupiter's Hill sphere is about 740 Jupiter radii in size.

Jupiter's largest and third-closest irregular satellite, Himalia, is 85 km in radius and was discovered as long ago as 1904, but most have been found by dedicated telescope surveys since the year 2000. The smallest are only about 1 km in radius. Even Himalia is only seven pixels across in the best Galileo image, so little is known of any of them.

Three other prograde irregular moons have orbits similar to Himalia (at about 160 Jupiter radii), and they all reflect sunlight in the same way as carbonaceous asteroids, so it is suggested that all four are fragments of a carbonaceous asteroid that broke up on capture by Jupiter.

Three groups defined by common reflectance characteristics and similar orbital radii, eccentricities, and inclinations are recognized among Jupiter's retrograde irregular satellites. They are each named after their largest member: the seven-strong Ananke group has orbits near 297 Jupiter radii, the thirteen-strong Carme group near 327 Jupiter radii, and the seven-strong Pasiphae group near 330 Jupiter radii. These are believed to be fragments of asteroids of other types. Many of Jupiter's irregular satellites, including more than twenty beyond the Pasiphae group, have no known families. Each could be either a small captured asteroid or comet nucleus.

Quite when these moons were captured is unknown. It would have been much easier to achieve very soon after Jupiter had formed, because it could then have had an extended diffuse atmosphere to provide the necessary drag to slow the incoming objects down sufficiently to become captured. Nor is it clear when the parent objects for each related group broke up. It could have been during the capture process, or afterwards because of a collision.

None of Jupiter's irregular satellites shows synchronous rotation. They are too small and too far from Jupiter for tidal forces to be effective, and this is borne out by the few examples where rotation periods have been measured and shown to be only a few hours, in contrast to orbital periods of hundreds of days. On the other hand they feel the Sun's pull so strongly that the shapes and inclinations of their orbits can vary markedly in only a few years.

The irregular satellites of the other giant planets follow a similar pattern to Jupiter's, and their origins are probably similar. The most distant moons are all retrograde, but closer to the planet prograde and retrograde moons are intermingled rather than being neatly segregated like they are at Jupiter.

Saturn has thirty-nine known irregular moons counting Hyperion (Figure 11), a 180 × 133 × 103 km radius moon that has an orbit

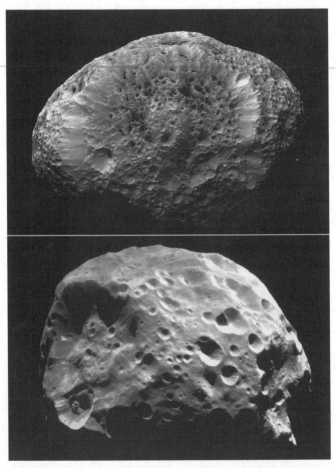

11. Hyperion (above) and Phoebe (below) seen at similar scales.

between those of the outer two regular satellites, Titan and Iapetus. Hyperion is unique among known moons in that its rotation is chaotic. Not only is its rotation period variable, but even its axis of rotation changes as it tumbles along. Its density is only half that of solid ice, and it probably has a porous, rubbly interior. It has a low albedo, suggesting a surface dusting of dark particles which is typical of this part of the Saturnian system.

Phoebe (Figure 11) is 109 × 109 × 102 km in radius and is the largest and closest of Saturn's retrograde irregular moons. Phoebe's orbit lies at 548 Saturn radii. This is too far out to be visited by Cassini after it had achieved orbit, so Cassini's approach to Saturn was timed to enable it to make a close pass by Phoebe on its way in, at a range of 2,000 km, making it the best-imaged example of all its kind. Cassini revealed a cratered surface and detected water ice, carbon dioxide ice, and clay minerals. Phoebe has an extremely low albedo, only 0.06, which may be because methane ice has been stripped of some of its hydrogen (long exposure to the Sun's ultraviolet radiation can do this), allowing the carbon atoms to link together as a black tarry goo. Phoebe is a good candidate to be a captured centaur—a class of icy asteroid found mostly beyond the orbit of Saturn. In 2009 infrared telescope observations revealed that Phoebe orbits within a diffuse but very broad (twenty times thicker than Saturn itself) belt of dust, thought to have been knocked off Phoebe's surface by micrometeorite impacts.

Saturn has two orbital groups of prograde irregular satellites. The individuals in one are given Inuit names such as Siarnaq, and in the other Gallic names such as Albiorix. Each group could be the remains of a larger moon destroyed by collision. Apart from Phoebe, Saturn's retrograde irregular satellites have Norse names, and include groupings that could each represent fragments of the same captured asteroid.

Irregular moons of Uranus and Neptune are challenging objects even for the best modern telescopes. Uranus has nine, discovered in the period 1997–2003. They have retrograde orbits except for Margaret, which has the most eccentric orbit of any planet's moon. The largest is Sycorax, about 75 km in radius, whereas the smallest known have radii of about 10 km. There are no close orbital groupings, and each is probably an individually captured object.

Only six irregular moons of Neptune are known: three prograde and three retrograde. The largest, Nereid, has a radius of 170 km. It was discovered in 1949, and the others in 2002–3. The outermost examples, Psamathe (20 km radius) and Neso (30 km radius), are in eccentric retrograde orbits at mean distances of 1,885 and 1,954 Neptune radii. This is a vast distance (they take over 9,000 days to orbit the planet), but these orbits are stable because Neptune's Hill sphere is larger than Jupiter's, thanks to its greater distance from the Sun.

Nereid may be a large surviving remnant of a regular satellite that was catastrophically destroyed (maybe in the Triton capture event). It has been shown to have water ice on its surface, with an imposed low albedo thanks to a darkening agent such as carbon, in which respect it resembles some of the regular satellites of Uranus. It has been proposed that one other irregular satellite, Halimede, could be a smaller fragment from the same body, but Neptune's other irregular moons are probably individually captured objects.

Inner moonlets

Inner moonlets can be very small, as their name implies, and their proximity to the glare of their planet makes them harder to detect by telescope than irregular satellites. There is a good reason why large moons are not found close to their planets, articulated by the French astronomer Édouard Roche (1820–83) who calculated the distance from each planet at which the difference between the planet's tidal pull on the moon's near and farsides would exceed the moon's own gravity. At this distance, commonly referred to as the 'Roche limit', a fluid or loosely consolidated body would be pulled apart, though the internal strength of a solid body allows it to approach closer before it disintegrates.

Most inner moonlets orbit within their planet's Roche limit, and are probably fragments of larger moons ripped apart by tides. Some of the more distant, and larger, examples may have originated as regular satellites battered by collisions.

Only four inner moonlets of Jupiter are known, of which only the largest, Amalthea, had been discovered before Voyager. The best Galileo images of these are included in Figure 12.

Saturn has eight known inner moonlets within the orbit of Mimas, and there are three others (only about 1 km in radius) between Mimas and Enceladus. Of those three, only Methone has been seen at close quarters by Cassini, and has a surprisingly smooth egg shape. The best Cassini images of Saturn's conventional inner moonlets are included in Figure 12, and most show much better detail than provided by Galileo at Jupiter. Saturn's closest inner moonlets, Pan and Atlas, have ridges running round their equators, giving them curious 'flying saucer' shapes. Their orbits follow gaps in Saturn's ring system, and the equatorial ridges are probably swept up ring material.

Janus and Epimetheus are a unique pair: the orbit of one is only a few km bigger than the other. When the inner, faster travelling, moon catches up with the other, which happens about every four years, their mutual gravitational interaction causes them to swap orbits. The previously faster one is now in the wider orbit so it slows down, whereas the previously slower one speeds up until it catches up with its fellow four years later and the cycle repeats.

Uranus has thirteen known inner moons, of which only Puck (Figure 12) was seen in any detail by Voyager 2, which discovered

12. A compilation of the best-imaged inner moonlets, shown at approximately correct relative scales (Pandora is 100 km from end to end). For Jupiter (Galileo images) and Saturn (Cassini images), moonlets are arranged with closest to the left. Two views of Atlas are included: equatorial (above) and southern hemisphere (below). Puck and Proteus images are from Voyager 2.

it. They are all dark (low albedo) objects, probably as a result of radiation damage to methane. Most were discovered by Voyager 2, but three (including the smallest, Cupid and Mab, which are only about 5 km in radius) are more recent telescopic discoveries. All thirteen are crowded into a narrow orbital range, 1.95 to 3.82

Uranus radii, and simulations suggest that they can disturb each other's orbits so that there is likely to be a collision sometime in the next 100 million years.

Six inner moonlets of Neptune were revealed on Voyager 2 images, and a seventh (the smallest, at about 18 km radius) was discovered by imaging with the Hubble Space Telescope. Like their equivalents at Uranus, they have low albedo, and seem to be water ice made dark by radiation-damaged methane. One of these, Proteus (Figure 12), is the largest inner moonlet in the Solar System, at $220 \times 208 \times 202$ km radius. Its shape is a long way from hydrostatic equilibrium and probably results from collisional battering. The innermost of Neptune's moonlets, Naiad, orbits well inside the Roche limit, and its eventual fate will probably be either to spiral into the planet's atmosphere or to be ripped apart by tides and make a new ring.

Rings and shepherd moons

Saturn is famous for its spectacular ring system. This was first seen by Galileo, but he couldn't make it out clearly. Huygens (in 1655) was the first to correctly interpret what he saw. The other giant planets have rings too, though these contain much less mass and so are far less spectacular.

Even Saturn's glorious rings contain less mass than Mimas, its smallest regular satellite. They are probably very old, and represent mostly either material that was too close to the young planet to have ever been gathered into a moon, or the remains of a moon that was ripped apart when it strayed within the Roche limit. The easily visible extent of the rings (which you can see for yourself with a small telescope) consists, working outwards, of rings designated C, B, and A by the IAU. These are only a hundred metres thick and extend from about 75,000 km to about 137,000 km from the planet's centre. That's 1.24 to 2.27 Saturn radii, but of course only 0.24 to 1.27 Saturn radii above the cloud tops. Spectroscopy shows that these rings are mostly water ice, darkened and in most places reddened by radiation damage or dusty contaminants. The relatively slow cooling rate of the rings when they pass into Saturn's shadow, the ability of ground-based radar to record a signal reflected from the rings, and the effect of the rings on Voyager 1's radio transmissions show that the rings are made of chunks ranging from about one centimetre to five metres in size. Each such chunk is in orbit about the planet. It would be perverse to regard every one of them as a moon, though there is no agreed lower size limit for what can be called a moon.

Less substantial rings occur to either side of the main rings. The D ring (66,900–74,500 km) lies between the C ring and the planet, whereas other rings lie beyond the A ring. Saturn's rings have amazingly complex internal structures, most obviously at distances from Saturn where the ring material becomes more diffuse or even vanishes completely. The widest 'gap' is 4,800 km wide and separates the A and B rings. It was discovered in 1675 by Giovanni Cassini and known as the Cassini Division. Its inner edge occurs at the distance where any ring particles would have twice the orbital period of Mimas, and so is readily explicable. Fine structure within the gap, at radii where particles are either concentrated or dispersed, is not fully understood.

Numerous minor gaps elsewhere in the rings defy simple orbital resonance explanation, but some are swept clear by inner moonlets whose orbits actually follow the gaps—for example, Pan and Daphnis each sweep a different gap in the A ring (Figure 13).

Cassini images in 2013 showed a 1,200 km long, 10 km wide enhancement of material at the outer edge of the A ring, which might be concentrated around a small mass, perhaps 1 km across. This could be a new moonlet in the making, but is more likely a temporary disturbance that will disperse.

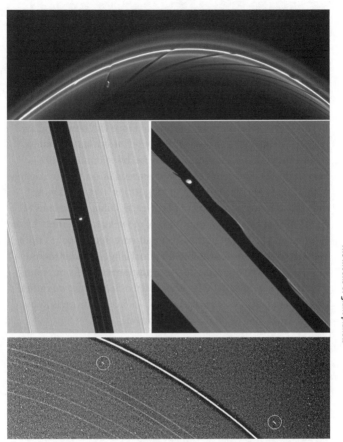

13. **Moons among the rings.** Top: Prometheus, 176 km long, travelling from right to left, making periodic disturbances in Saturn's F ring. Middle left: Pan, 34 km long, orbiting in the Encke gap, casting long shadow thanks to grazing-incidence sunlight. Middle right: Daphnis, 8 km long, orbiting in the Keeler gap, making waves in the inner edge of the A ring. Bottom: Uranus' rings with two moons circled, Cordelia (left) and Ophelia (right). Each is about 20 km in radius, and they appear artificially elongated in this long-exposure image. There is a lot of speckly noise, and the dark band either side of the outermost ring is an artefact.

The F ring is only 30–500 km wide and lies outside the A ring. This is held in place, and given a dynamic structure, by the inner moonlets Prometheus (Figure 13) and Pandora that orbit just inside and just outside it, respectively. Inner moonlets such as these, whose intimate association with a ring system helps to define its shape, are often referred to as 'shepherd moons'.

Beyond the F ring there are various tenuous rings and ring arcs believed to be dust knocked off the outermost inner moonlets, Aegaeon, Methone, Anthe, and Pallene, by micrometeorite bombardment, and confined or concentrated into arcs by resonance with Mimas. Further out still, at 180,000 to 480,000 km, is the diffuse E ring consisting of particles less than a micrometre across made of water ice and traces of other material erupted from Enceladus. The belt of diffuse dust that envelopes Phoebe's orbit is sometimes referred to as an additional ring, but is far too thick for the term to be appropriate. A suggestion dating from 2008 that Saturn's regular satellite Rhea is encircled by its own diffuse ring system has now been discounted. In fact, no moon is known to have rings, just as no moon has a moon.

Jupiter's rings were discovered by Voyager 1. The main ring is only 6,500 km wide and 30–300 km thick. The orbit of the innermost inner moonlet, Metis, occupies a gap within this ring, and the ring's outer edge is shepherded by the orbit of Adrastea. The ring material is an unknown reddish material, consisting of a mixture of micrometre-sized dust and larger chunks, with a total mass no more than about five times that of Adrastea and possibly much less. It would take less than 1,000 years for radiation pressure and interactions with Jupiter's magnetic field to disperse the dust, so unless it is extraordinarily young it must be being replenished either by collisions between the larger chunks, or by micrometeorite bombardment of Metis and Adrastea.

NASA's New Horizons mission searched in vain for unknown inner moonlets within this ring during its 2007 fly-by, so there are unlikely to be any greater than 0.3 km in radius. It did however find seven clumps of material, occupying arcs extending for 1,000 to 3,000 km.

Inward of the main ring, Jupiter has a 12,500 km thick 'halo ring', which seems to be dust spiralling inwards to Jupiter from the main ring. There are also two exceedingly faint 'gossamer' rings, which are dusty features extending outwards from the orbits of Amalthea and Thebe, the outermost inner moonlets, which are probably supplied by ejecta from micrometeorite bombardment of their surfaces.

The rings of Uranus were discovered in 1977, by telescopic observers who noted the repeated dimming of a star as the otherwise invisible rings passed across it. Most of our knowledge of the rings comes from Voyager 2 and the Hubble Space Telescope. Thirteen rings are now known, with a total mass exceeding Jupiter's rings but much less than Saturn's. Several are narrow and consist mostly of boulder-sized chunks of low-albedo reddish material, believed to be produced by the fragmentation of inner moonlets. These are all narrow rings, the outer and brightest of which (the epsilon ring) is shepherded by the moonlets Cordelia and Ophelia (Figure 13). It is at a mean distance of 51,150 km from the planet's centre, but is eccentric in shape so that the distance varies by 800 km around its circumference. It is 20 km wide where closest to the planet and 100 km wide where furthest. The inner rings are even more tightly confined but have no known shepherd moons, prompting the suggestion that these formed by fragmentation no more than about 600 million years ago.

Uranus' other rings are dusty. There are four inside the narrow rings, believed to be short-lived dust spiralling towards the planet, and two beyond—the outermost of which is spread round the orbit of the outermost inner moonlet Mab, which is probably the source of its dust.

Neptune has five rings, which are extremely dark, like those of Uranus. They are about half dust and half larger chunks. Material in the outermost ring is concentrated into several arcs. Attempts to explain this as a 42:43 resonance with the inner moonlet and ring shepherd Galatea have failed, and it is clear that there is much about rings and their relationship with moons that we do not yet understand.

Chapter 5
Regular satellites in close up

Twenty-five years ago I wrote a Voyager-based book describing the regular satellites of the giant planets as 'worlds in their own right'. The truth of that phrase is even more evident now. Each of them is individually fascinating, and two are even widely regarded as better candidates than Mars for hosting extraterrestrial life. Here I will discuss some of my favourites.

Io

Thanks to tidal heating, Io is the most volcanically active body in the Solar System, surpassing even the Earth. Nine eruption plumes were discovered by the Voyager probes, of which the largest was 300 km high and more than 1,000 km wide. The Voyagers detected strong infrared radiation from the plume sources demonstrating high temperatures, and increasingly sophisticated infrared telescopes on Earth documented these and many other 'hot spots' on Io during the sixteen-year interlude between Voyager and Galileo, and in the periods before and after the 2007 fly-by by New Horizons.

The discovery of active volcanoes on Io led the IAU to revise the naming convention that had been lined up for features on Io, and to choose instead names of mythical blacksmiths and fire, sun, thunder, and volcano gods and heroes. For example, of the erupting volcanoes in Figure 14, Prometheus is a Greek god who stole fire to give to humans, Tvashtar is named after a sun god in Sanskrit epics, and Masubi is a Japanese volcano deity. Eruptions have now been documented at about a hundred sites, of which a few (such as Prometheus) are persistent and some others (such as Tvashtar and Masubi) have erupted on several occasions.

Io lacks any visible impact craters, because it is continually being resurfaced by volcanism that rapidly buries craters. This resurfacing occurs by a combination of fallout from eruption plumes and over-riding/flooding by lava flows. The strong yellow and orange colours that dominate Io's surface were originally thought to show that many or all of the lava flows were made of sulfur, but infrared measurements of the active vents show that the temperature within them is far too hot for sulfur. The hot material must be molten rock, probably a low-viscosity magma low in silica and rich in magnesium.

Low-albedo lava flows are recognizable elongated lobate features up to 300 km long, and often emerge from caldera-like landforms whose circumferences are neither smooth nor circular enough to be impact craters. After a lava flow has ceased to move, it begins to fade from visibility as it becomes coated with sulfur and sulfur dioxide frosts (responsible for Io's colouring), which are both exhaled from the ground and deposited as fallout from eruption plumes.

Some calderas have no flows emerging from them but are strong sources of infrared radiation, and evidently contain 'lava lakes' where magma continually wells up, crusts over, and sinks to be replaced by fresh magma from below.

Other volcanic vents or fissures are the sources of Io's largest eruption plumes. These are presumably powered by the expansion of gas bubbles in magma as it rises through the crust, in which the gas expands with sufficient violence to shatter the magma to

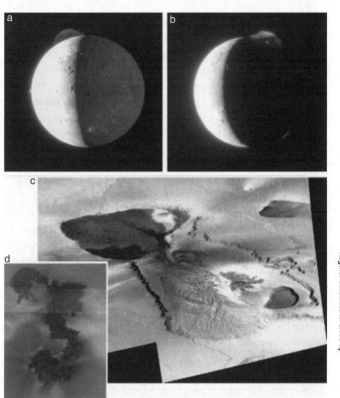

14. (a) and (b) Io imaged by New Horizons on 28 February and 1 March 2007. The 290 km high eruption plume from Tvashtar is prominent in the north. The 28 February image also shows a 60 km plume from Prometheus on the limb in the nine o'clock position, and the top of a third plume, from Masubi, rises high enough to catch the sunlight in the south. The nightside of Io is lit by light reflected from Jupiter, but the dayside has been overexposed to increase the plumes' visibility. On the 1 March image, the vent at Tvashtar can be seen glowing in the dark. The Masubi plume can be seen above the dark limb in the four o'clock position; c 300 km wide mosaic of Galileo images of Tvashtar as it was on 16 October 2001. Shadows change confusingly because images from different times of day have been combined; d 90 km wide mosaic of Galileo images of lava flows erupted from a caldera (at top left). The Prometheus plume originates near the front of the main lava flow, which is towards the bottom of the image.

fragments and propel it skywards. On Earth, this process would draw in the surrounding air and develop into a convecting column, but Io has virtually no atmosphere so the particles follow parabolic arcs taking them skywards and then back down to the ground. This is well demonstrated by the Tvashtar plume in Figures 14(a) and 14(b), where the eruption speed was about 1 km per second. The gas that drives the eruption is a mixture of sulfur and sulfur dioxide. Both substances condense into 'frost' particles near the top of their trajectories, and fall to the ground along with the fragmented magma, but in this case the magma dominates so that the deposit has a low albedo.

The Prometheus and Masubi plumes are a different sort. In this case, the plume does not originate at a volcanic vent but instead at the advancing front of a lava flow. In this situation, the hot lava vaporizes the sulfur dioxide surface-frost of the over-ridden surface. It is this that forms the plume, which becomes visible as the expanding gas condenses to frost particles. Plumes of this kind are mainly sulfur dioxide with little or no magma, and form a high-albedo deposit where they fall. Many such white blotches are visible in Figure 14(a), especially in the Jupiter-lit nightside area.

An average of about one tonne per second of oxygen and sulfur that has been erupted from Io is ionized by interaction with Jupiter's magnetic field (which envelops Io), and feeds a doughnut-shaped 'plasma torus' that encircles the planet at the radius of Io's orbit.

Europa

Europa could scarcely look more different to Io (Figure 15), though if we could strip away about 100 km of ice and water we might reveal a less active version of Io beneath. Europa's surface is high-albedo ice, too young to have accumulated more than a few impact craters. Much of the surface is a mass of ridges and grooves,

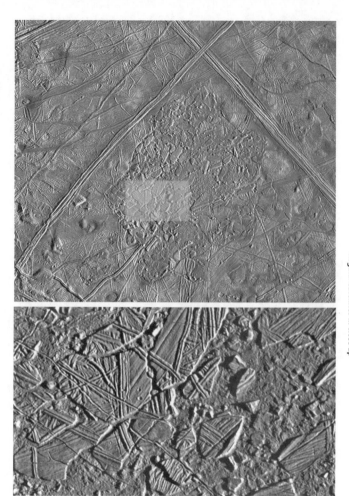

15. Europa imaged by Galileo. The upper image is a 300 km wide area, showing ball of string terrain, except centrally where it has been broken apart to form chaos terrain. The area in the transparent rectangle is shown below at higher resolution, revealing rafts and matrix. The two largest impact craters in the lower image are less than 200 metres in diameter, just left of centre.

contributing to an appearance that has been likened to a ball of string. Most ridges seem to be where a crack has opened and then closed again, squeezing out some slush to make a ridge at the surface. This opening and closing could be a tidal process, and so would repeat with each eighty-five-hour-long orbit round Jupiter, building a hundred-metre-high ridge of refrozen slush over a crack's active lifetime before it seals for good and a new crack opens elsewhere. Some particularly large ridges are fringed by discoloured ice, demonstrating impurities (probably magnesium sulfate and other salts) in the water from which the slush froze.

The ball of string terrain is disrupted in various regions, mostly by being broken apart to form a landscape described as 'chaos'. Here, you can see rafts with ball of string textures separated from each other by a lower-lying surface that has a fine jumbled texture. Most scientists now accept that this 'matrix' between the rafts is the refrozen surface of a body of water that had been temporarily exposed when the surface ice became thin, allowing it to break into rafts that drifted apart. In some chaos regions it is possible to see how the rafts once fitted together, but in others whole rafts have vanished—either sunk or melted. In older chaos regions, some new ridges and grooves cross rafts and matrix alike, suggesting that eventually chaos becomes unrecognizably overprinted into a new generation of ball of string terrain.

Even very salty water would freeze quickly upon exposure to space at Europa's −160°C surface temperature, but we can imagine kilometre-thick ice rafts drifting through an ocean surfaced by tens of metres of slush for many weeks before the slush became too thick and stiff to permit further drifting. Calculations based on the typical hundred-metre height difference between raft surface and matrix surface, and allowing for the likely buoyancy difference between raft and salty water, suggest that when the ice sheet broke into rafts it was between half a kilometre and a few kilometres thick.

One thing that experts don't yet agree on is whether the body of water exposed when each chaos forms is the top of Europa's subsurface global ocean (Figure 9), or merely the top of an isolated lens of liquid water within Europa's ice shell. We have no way of telling the thickness of this ice shell (a Europa orbiter to map the height of the tidal bulge and equipped with ice-penetrating radar would be the best way). The shell could be tens of kilometres thick, in which case the melt-through process to thin the shell upwards from its base until it was thin enough to break apart would require a sustained localized input of heat capable of melting tens of kilometres of ice, whereas if the shell is only a few kilometres thick melt-through would be much easier to achieve. The heat to drive this process would be tidal heat generated in Europa's rock interior, and possibly delivered to the floor of the ocean by a submarine volcanic eruption.

Proponents of the thick-ice model of Europa prefer a temporary lens of liquid water to be melted within the ice, by heat transported upwards by solid-state convection within the deeper ice. This could either be powered by tidal heating within the ice, or be a response to a submarine volcanic eruption conveying heat to the ocean.

Even though it has essentially no atmosphere, Europa is a more promising, though less convenient, place to search for extraterrestrial life even than Mars. Life on Earth is believed to have begun at 'hydrothermal vents' on the ocean floor, and the tidal heating in Europa's rocky shell that supplies the heat to keep the subsurface ocean from freezing should be associated with hydrothermal vents aplenty, where water that has been drawn inwards through the rock is expelled after heating.

There are ecosystems in Earth's deep oceans that survive on the chemical energy emerging from hydrothermal vents. These do not depend at all on sunlight, which plants use to power their metabolism. It is entirely feasible that something similar exists round vents on Europa's ocean floor, where they might be 'chemosynthetic' microbes at the base of the food chain, and larger, possibly multicellular, predators feeding on them.

If life began at deep-sea vents on Europa, it may have subsequently adapted to colonize other environments that would be easier for us to explore. The best would be a tidal crack—up which water would have been drawn each time it had opened. Any planktonic organisms that reached the upper few metres would find enough sunlight to allow photosynthesis, and provided that they stayed at least a few centimetres below the surface the water would shield them from the severe radiation dose that would be experienced at the surface. Some unlucky organisms would be squeezed out with the slush each time a crack closed, so the easiest place to find fossilized life would be in entombed in the frozen icy slush that makes up a surface ridge.

Ganymede and Callisto

Ganymede may also have an ocean resting on rock, which is an important factor for the supply of chemical energy for any life to feed on. As suggested in Figure 9, this may be overlain by several successive solid and liquid shells. However, whatever the exact situation, the top of even the topmost ocean is deep down and remote from the surface. Ganymede has no recognizable traces of chaos terrain, so maybe its surface ice shell never became thin enough for melt-through to occur. However, there have been multiple episodes in which parallel swarms of grooves were formed. The youngest examples, which cross-cut older terrain, have higher albedos than other areas (Figure 16). This might be Ganymede's local equivalent of ball of string terrain, but most grooves are cracks without the ridges seen on Europa.

An important contrast with Europa is that Ganymede's surface fracturing seems to have ceased long ago. Figure 16 shows numerous craters from 2 km downwards on the belt of high-albedo, youngest

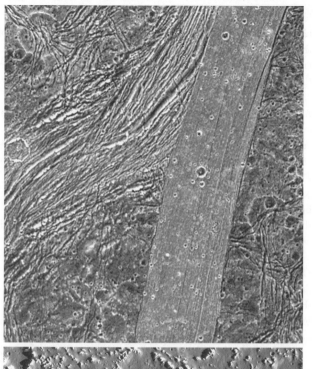

16. Representative 70 km wide regions of Ganymede (top) and Callisto (bottom), imaged by Galileo. The youngest terrain on the Ganymede image is the high-albedo belt of fine grooves running north–south, but even this must be very old because there has been time for many impact craters to be made on it. Note that the Sun was high in the west for the Ganymede image, but lower and in the east for the Callisto image.

terrain. The older, darker terrain has craters like this too, but it also has craters up to 10 km in size that have been cut by many of the cracks. Beyond the region shown in Figure 16, Ganymede has craters up to 200 km in size, some of which are superimposed on high-albedo belts.

Although the crater density on Ganymede's surface is convincing evidence that the surface is old, sadly we cannot apply the lunar cratering timescale to derive an absolute age. This is because if you plot the distribution of crater frequency versus crater size for Ganymede, it looks very different to the equivalent plot for the Moon. This shows that Ganymede has been struck by a different population of impactors, and that there is no reason to assume that the history and rate of bombardment at Jupiter have been the same as at the Earth, five times closer to the Sun.

However, we *can* conclude that the dramatic contrast in crater densities between Ganymede and Europa provides clear evidence that the average age of Ganymede's surface is much older than Europa's. It could easily be older than three billion years, whereas no surviving surface on Europa is likely to be older than about fifty million years.

Callisto has an even more heavily cratered surface than Ganymede, and if it ever did have surface cracks these have been obliterated by the cratering. Figure 16 shows an area of Callisto at similar scale to the accompanying view of Ganymede, with craters up to 15 km wide, but there are plenty of bigger craters elsewhere. In fact Callisto has the largest impact crater in the Solar System, a 3,800 km diameter multiple-ringed basin named Valhalla. Although Valhalla's rings have survived, it no longer retains its original depth and has become flattened because Callisto's ice was unable to sustain such a deep basin, and flowed upwards to repair the damage. Callisto's internal ocean (Figure 9) is inferred from its induced magnetic field. It has had no discernible effect of the surface and must lie much deeper than Europa's ocean.

Titan

Although Saturn has seven regular satellites, Titan is the only one to rival Jupiter's Galilean moons for size. It is only slightly smaller and less massive than Ganymede (see Appendix), and has an icy outer shell, which Cassini observations showed to be separated from the interior by an internal ocean beginning at a depth of about 100 km. If this ocean is water mixed with ammonia it could be about 200 km thick, and below it would be a 200 km thick layer of Ice VI surrounding Titan's rocky interior.

The thing that makes Titan particularly special is that, unique among moons, it has a dense atmosphere, whose surface pressure is about one and a half times that on Earth. It is about 98 per cent nitrogen with most of the rest being methane (CH_4). Between 50 and 250 km above ground level, solar ultraviolet light splits hydrogen atoms off from methane (a processes called photodissociation), whereupon the carbons link into chains to make a hydrocarbon haze that hides the surface from view at most wavelengths of light. The Voyagers saw nothing of the surface, but Cassini carried an imaging radar specifically to penetrate this haze and map the terrain. In addition, Cassini's optical camera was able to obtain low-resolution views of the surface using a special filter, which has proved useful for tracking seasonal surface changes. We have also seen the surface from the Huygens probe (carried to Titan by Cassini), which sent back clear pictures from below the haze during its parachute descent and from the ground itself.

And what a surface was revealed! If you did not know that the surface 'rock' is actually water ice, and the few fluffy low altitude clouds are condensed methane and/or ethane, you might think you were looking at a part of the Earth. Titan experiences methane rainfall, which collects in streams and rivers to erode valleys, which drain into lakes, three of which are large enough to be called seas, such as Kraken Mare in Figure 17. These seas and most of the lakes are at high northern latitudes, but there are lakes

17. Map-projected 450 km wide area near Titan's north pole, compiled from Cassini imaging radar swaths of variable quality collected in the years 2004–13. North is to the left. The 80° N line of latitude is superimposed, plus lines of longitude at 10° intervals. The large dark area is Ligeia Mare, less than five metres deep near its northern shore allowing radar to reflect from the seabed, but in excess of 100 metres deep elsewhere. Several rivers can be seen draining into it.

near the south pole too. Dry lakebeds have also been identified, with some signs of seasonal changes. Evidence of changing lake volumes comes also from the characteristics of their shorelines; for example, much of the southern shore of Ligeia Mare looks like a typical drowned coastline where hills and valleys carved on dry land have been flooded by a rise in sea level. There is also the possibility of spring-fed ponds and lakes. The liquid seeping into them would have begun as methane rain that seeped into the icy crust and percolated underground (through what on Earth would be called an 'aquifer'), where some of it could be converted to ethane or propane before emerging again.

The Huygens lander touched down only a little way south of the equator. It saw dry river valleys on its way down, and landed on a flood plain or dry lakebed. The area was strewn with pebbles of ice that had been rounded by transport along a methane river, sitting on a dark substrate that could be a tarry mix of rained-out hydrocarbons.

Vast fields of low-albedo sand dunes cover the icy 'bedrock' at low latitudes, with individual dunes about a hundred metres high and tens of hundreds of kilometres long. These dunes appear to be formed parallel to eastward-blowing winds, driven by tidal forces from Saturn. The 'sand' grains in these dunes could be dirty grains of ice, or hydrocarbon particles.

Despite the depth and density of Titan's atmosphere, large impactors should be able to reach its surface at the hypervelocities needed to make impact craters, but only eight impact craters have been identified, ranging from 29 to 292 km in diameter. This paucity of craters shows that the average surface age is young, but we do not know whether the older craters have merely been erased by erosion (rivers and wind) and buried by sediment, or whether other processes are at work too. One of these could be volcanism. There are a few candidates for volcanoes on Titan, such as Sotra Patera, a twin-peaked mountain 1.5 km high, with craters from which apparent lava flows have emerged. The lavas in this case would not have been molten rock, but something derived by the partial melting of Titan's icy crust. It could have been an ammonia-water mix (which is liquid at a much lower temperature than pure water), or even a waxy material derived by processing buried hydrocarbons that originally rained out of the atmosphere.

To distinguish it from conventional volcanism involving molten rock, as on Earth, the Moon, and Io, volcanism in which the erupted material is derived from ice of any kind is called cryovolcanism. On Titan, significant tidal heating seems unlikely, so the driver may be heat from radioactive decay in the rocky interior.

Enceladus

Enceladus is the smallest of Saturn's regular satellites apart from its inner neighbour Mimas. However, whereas Mimas is covered in impact craters, Enceladus is a much more interesting place. Voyager showed cratered terrain juxtaposed against terrain that looked smooth and featureless. The higher-resolution imaging achieved by Cassini revealed that the apparently smooth areas are in fact intensely fractured, and that the impact craters that used to be there have been largely erased by multiple generations of cross-cutting fractures.

Even more exciting was Cassini's discovery of plumes being erupted from a series of fissures near the south pole (Figure 18), that became colloquially known as the tiger stripes because the surface along them shows up as bands of slightly bluer, fresher ice. More than a hundred individual geysers have been identified, which jet tiny crystals of water ice into space as speeds of about 1 km/s. These are evidently the source of the material forming Saturn's E ring. The plumes were not anticipated when Cassini was designed, but the mission plan was revised to allow the craft to swoop through them at ranges of 200 to 25 km above the surface. Cassini's Ion and Neutral Mass Spectrometer (which had been intended primarily as a tool for sampling Titan's exosphere

18. **Enceladus seen by Cassini on 21 November 2009 (top); south polar region imaged by Cassini on 30 November 2010 (bottom), showing sunlit plumes emerging from three sub-parallel fissures.**

and ions, atoms, and molecules associated with Saturn's magnetosphere) showed that the plume is 99 per cent water with traces of methane, ammonia, carbon monoxide, carbon dioxide, and various simple organic molecules. Cassini's infrared spectrometer was able to show that the temperature inside the ten-metre-wide fissures from which plumes erupt is at least as warm as $-70°C$, in contrast to the regional surface temperature of $-180°C$. Enceladus' orbit is only very slightly elliptical, and the plumes are strongest when Enceladus is at its furthest from Saturn, which is when the stresses in the crust would tend to pull the tiger stripe fissures open.

The source of the vented material is thought to be a vast pod of melted water below the south pole, warmed by tidal heating. Tidal cracking is presumably responsible for the vastly complex history of surface fracturing exhibited across the globe. It is far from certain that the present-day rate of tidal heating is sufficient to sustain the current activity, which may be left over from a previous episode of stronger heating.

Unlike Europa, Enceladus probably does not have a global ocean and nor is the liquid water pod likely to be resting on rock. The small extent and the limited chemical interaction between water and rock means that the potential for Enceladus to host microbial life is less than for Europa. However, the delivery of samples to space where they can be accessed without landing means that a future fly-by or orbiter probe with specially designed instruments could search for organic molecules or biologically driven fractionation of isotopes, without needing to land on the surface.

Before Enceladus' plumes had been discovered, the Galileo mission searched for eruption plumes from Europa from 1995 to 2003 without success, and nor were any seen during Cassini's 2001 fly-by of Jupiter. In 2012 the Hubble Space Telescope detected a diffuse zone of atomic oxygen to one side of Europa that is most simply explained as a product of photodissociation of water vented from a plume. However, repeat attempts at detection throughout 2014 were blank, and if eruption plumes do indeed occur at Europa it seems clear that they must be weaker and less persistent than those currently occurring at Enceladus.

Irrespective of whether or not there is life in the liquid water zones of Europa and Enceladus, it is conceivable that the right kind of terrestrial microbes could survive and reproduce if they found themselves there. Space probes are cleaned and sterilized before launch, but it is impossible to remove or kill all the microbes, and some can survive for years in space. If probes such as Galileo and Cassini were left as derelict hulks in orbit, there is a chance that one day they might crash into Europa or Enceladus, unintentionally delivering viable microbes from Earth. To prevent this, at the end of its life in 2003 Galileo was deliberately crashed into Jupiter, and Cassini will meet a similar fate at Saturn in 2017.

This is not just a matter of the ethics of protecting alien ecosystems, though arguably that should be an important consideration behind any so-called 'planetary protection' protocols. We also want to study any such ecosystems free of the confusion that any spacecraft stowaways would engender. If these ecosystems exist, it will be important to establish whether life began there independently of life on Earth, or whether life within the Solar System shares a common origin, perhaps having been spread from world to world as accidental passengers inside meteorites.

There can hardly be any bigger question than that of whether life started independently more than once in our Solar System. Until we find unrelated life somewhere, life on Earth could be just an extraordinarily rare statistical fluke. But if life started inside an icy moon independently of life on Earth, then surely it has also begun on other habitable worlds throughout our galaxy. A dedicated mission to seek life at Enceladus or Europa might be the best chance we have of revising the answer to the question 'Are we alone?' from 'maybe' to 'no!'

Iapetus

Iapetus orbits Saturn at almost three times the distance of Titan, making it the most distant of any giant planet's regular satellites. It is also unusual in that its orbit is inclined at 15.5° to Saturn's equator, as a result of which it is the only regular moon of Saturn from whose surface the planet's rings would sometimes be clearly visible.

When Giovanni Cassini discovered Iapetus in 1671 he was able to see it only when it lay to the west of Saturn in the sky. It took him more than thirty years of trying before he glimpsed it faintly when it lay east of Saturn. Cassini correctly interpreted this as proof that Iapetus is in synchronous rotation and that its forward-facing hemisphere has a much lower albedo than the trailing hemisphere.

Thanks to the Cassini spacecraft, we now know that the surface distribution of albedo resembles the pattern on a tennis ball (Figure 19). A high (0.5–0.6) albedo tract runs from pole to pole via the centre of the trailing hemisphere. Its low (0.03–0.05) albedo counterpart is a broad equatorial tract centred on the leading hemisphere, and has been named Cassini Regio in honour of Giovanni Cassini. All other features on Iapetus are named after characters and places in the medieval French epic poem *Chanson de Roland*, which tells of a conflict between Franks and Saracens. With the exception of Cassini Regio, names in the high-albedo tract are generally Frankish and names in the low-albedo tract are generally Saracen.

Iapetus is very much a two-toned world, lacking shades of grey even at the most detailed image resolution of about thirty metres. This dichotomy could have been kick-started by Iapetus' leading hemisphere sweeping up dust sourced from impacts on Phoebe or smaller irregular satellites, but the sharp edges of the main dark tract and of small patches in the transition zone between the two tracts show that this cannot be the whole story. The dark material is probably a carbon-rich 'lag deposit' less than half a metre thick that has accumulated from impurities within the ice that were left behind while surface ice has very slowly sublimed away into space (which means passed directly from solid to vapour without

19. Iapetus seen by Cassini. The leading hemisphere (top); the trailing hemisphere (bottom). The albedo contrast between bright and dark sides is actually more extreme than it appears here.

melting). Iapetus' slow (seventy-nine days) rotation means that its surface has longer to warm up while facing the Sun than is the case for any other, faster spinning, regular satellite of Saturn.

Darker surfaces absorb more sunlight and heat up more than reflective surfaces, so that once an albedo contrast is established it will get stronger over time, until the non-volatile dark lag deposit is so thick that ice can no longer sublime from below it. Currently, the equatorial noontime temperature reaches about −144°C in Cassini Regio, whereas it is about 16°C colder in the high-albedo terrain.

Iapetus has an ancient, heavily cratered surface. Dark material sits on 'warm' crater floors whereas cold, pole-facing, inner walls of craters remain bright in the transition zone between high- and low-albedo tracts. Other peculiarities of Iapetus reside in its shape. Its polar radius is 34 km less than its equatorial radius. Allowing for Iapetus' low density, this degree of polar flattening corresponds to the equilibrium shape of a body rotating in only ten hours, and so the shape seems to have been frozen at a time before tidal drag had forced the rotation to synchronize with the orbit.

In addition, Iapetus has a narrow equatorial ridge about 13 km high, with peaks of up to 20 km. It can be traced for 1,300 km throughout Cassini Regio, where it can be made out in Figure 19, but it is present only as isolated 10 km peaks in the high-albedo tract. It must be ancient, because it is overprinted by numerous impact craters, and its origin is a mystery. It may be a feature inherited from Iapetus' formerly rapid spin. Alternatively, it may have something in common with the equatorial ridges on Atlas and Pan, but in this case it could not have been gathered from Saturn's rings and would have to be material accreted on to Iapetus' surface from a long-vanished ring of its own.

Miranda

Beyond Saturn, we come to giant planet moons visited only by Voyager 2 during fly-bys. Mission planners had a particular problem in the Uranus fly-by. Uranus' axis of rotation, and with it the orbits of its regular satellites, is tilted at 98° to its orbit.

Technically this means that Uranus' rotation is retrograde. The moons orbit in the same direction that the planet spins, so they count as prograde. However, the important point is that in January 1986 when Voyager 2 flew past, Uranus' south pole was facing more or less towards the direction from which the craft arrived, so the moons' orbits presented themselves like the rings around a target. It was not possible, as had been the case at Jupiter and Saturn, to cross each moon's orbit in turn and to time the encounters so that as many moons as possible lay in parts of their orbits close to where the probe passed. Instead, all that could be done to maximize the science at Uranus was to aim for a close fly-by of the innermost moon currently known, and to make the best of distant images of the other moons. Science was also limited because the northern halves of each body were in long-term seasonal darkness, and could not be imaged at all.

The innermost known moon was Miranda. This only has a 234 km radius and lacks present-day orbital resonance, so it was expected to be a fairly dull, heavily cratered object. However, it turned out to be startlingly complex and an absolute joy to behold (Figure 20). About half of the sunlit hemisphere has a uniform albedo and is densely cratered, but the older craters have muted profiles as if a layer of drab material has been draped over them, whereas the younger craters have fresh, crisp profiles.

The remainder of the sunlit hemisphere is taken up by three separate regions of lineated texture and (in two out of three cases) varied albedo. From left to right in Figure 20, these are Arden Corona, Inverness Corona, and Elsinore Corona, which take their names from places in Shakespearean plays. The IAU-approved descriptor term 'corona' signifies an 'ovoid-shaped feature', but does not help us understand what they are. The coronae have fewer impact craters than elsewhere, and they are all fresh looking, so clearly the coronae have younger surfaces than the rest of the globe. There is quite a lot of fracturing associated with

20. Voyager 2 views of Miranda. A mosaic of the sunlit hemisphere (top); slightly rotated view of the cliffs at the lower edge of the other image, seen at 0.7 km per pixel from a range of 36,250 km (bottom).

Inverness Corona, which cuts into the adjacent terrain and has produced some impressive cliffs up to 10 km high.

An early theory that each corona is a reaccreted fragment of a previous moon that had been broken apart by collision now seems too naïve, and we have to construct a more complex story. Miranda was probably once a passive, densely cratered globe, parts of whose surface still survive below the mantling deposit. It experienced one or more episodes of tidal heating, during which fracturing and localized cryovolcanism produced the coronae. The coronae may be surfaced by icy magmas that welled up through fissures to produce the ridges that dominate much of their area. If there were explosive eruptions at the same time, this could explain the muted profiles of the pre-existing impact craters on the terrain beyond the coronae.

This all makes sense, at least superficially, though it may turn out to be wrong when we eventually see Miranda, especially the as-yet unimaged half, in more detail (no mission to Uranus is currently planned). Tidal heating could have been driven by an interval of 3:1 orbital resonance with Umbriel, or 5:3 resonance with Ariel.

Ammonia is much more effective than salts at depressing the temperature at which a watery fluid will freeze, and given the likely abundance of ammonia in ices at this distance from the Sun, the cryomagma erupted on Ariel is probably an ammonia-water mixture that can be melted out from ice at a temperature as low as −97°C. This is lower than the melting temperature of either pure water or pure ammonia in isolation, and is an example of how mixtures of ices mimic the behaviour of mixtures of silicate minerals in rock, where also the melting temperature of rock is lower than the melting temperature of separate minerals.

Of Uranus' other regular satellites, Voyager 2 showed hints of major fractures on Oberon and obvious fractures on Titania, but only on Ariel did it reveal clear evidence of both fracturing and cryovolcanism.

Ariel

Ariel is the second largest of Uranus' satellites, and Voyager 2 saw it in moderate detail from a range of 127,000 km during its approach, before switching its attention to Miranda. It is denser than any moon of Saturn apart from Titan and is probably about half ice and half rock (which, this far from the Sun, could be quite rich in carbon). In addition to water ice, spectroscopic studies from Earth have detected carbon dioxide ice, but ammonia (which has no easily detected spectroscopic signature) is likely to be there in abundance too, to judge from the nature of the cryovolcanic resurfacing that is apparent in many areas (Figure 21).

Ariel's surface has many fractures, bounding down-dropped strips of terrain displaced by a series of fault movements at different times. This sort of faulting is symptomatic of stretching of the crust, and may reflect an episode of internal heating (which could have been tidal, or, if early enough, radioactive heating) causing thermal expansion of the interior. These valleys are widest in the centre and lower right of Figure 21, where it can be seen that their floors have fewer craters than the high ground to either side, showing that they have been flooded by something. This 'something' seems to have been very viscous, judging from what appears to be the edge of a flow cutting across the crater that is indicated by the arrow in Figure 21. On the Moon, where the lava was far less viscous, the flow would have spread across the entire crater and completely flooded it. An ammonia-water melt would move in a sufficiently viscous fashion under Ariel's low surface gravity (less than a tenth of the Moon's) and would fit the bill perfectly here.

However, a feature that seems not to fit with very viscous flow is the sinuous channel running near the middle of the wide faulted valley near the lower right of Figure 21. This is reminiscent of Hadley Rille

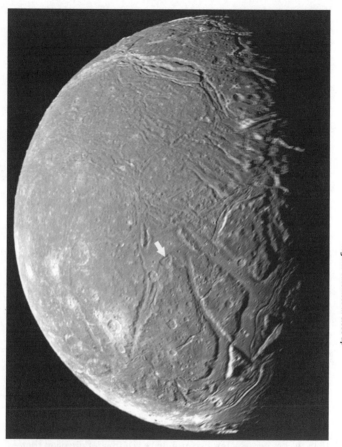

21. Voyager 2 view of Ariel, from about 130,000 km. The arrow marks
an impact crater that has been partly flooded by a viscous cryovolcanic
flow coming out of the faulted valley below and to the left. A slightly
smaller adjacent crater is superimposed on this flow, and so must be
younger.

(Figure 6) and many similar features on the Moon, whose origin is
accepted as thermal erosion by fast-flowing lava or collapse of the
roof of a lava tube—both of which require low viscosity.

There is so much we don't understand about Ariel, which has had a complex history, and which we are unlikely to sort out in my lifetime. How many different kinds of cryomagma were erupted on Ariel? Was there any sideways slip between the faulted blocks of terrain (there are hints of this if you look at the lower right of Figure 20)? What are those mountains silhouetted against the blackness of space at the upper left?

Triton

In many ways, Voyager 2 saved its best until last. It passed within 25,000 km of Triton, imaging about 40 per cent of the surface and revealing a complex surface that is rich in a diversity of ices, with a polar cap of nitrogen ice (Figure 22). I remarked on Triton's likely origin as a captured Kuiper belt object in Chapter 4. However although its composition is similar to Pluto, it differs markedly from Pluto in terms of landscapes and surface history.

Triton has a nitrogen atmosphere with traces of methane, but it is far less substantial than Titan's atmosphere and the surface pressure is only a few hundred thousandths of the Earth's. Tenuous though this is, it is sufficient to support a few wispy cirrus-like clouds of nitrogen ice particles a few kilometres above the surface and a 30 km high hydrocarbon haze layer. Triton has a global average albedo of 0.78, which is exceeded among regular satellites only by that of Enceladus. This is not so much because the surface is young (nearly 200 impact craters have been identified, all less than 30 km across), but because there is a frost of nitrogen that freezes out of the atmosphere to coat the −230°C surface.

Triton has peculiar seasons. Although its own spin is perpendicular to its orbit about Neptune, this orbit is inclined at 157° to the planet's equator (so it is retrograde). On top of this, Neptune's equator is tilted at 30° relative to its orbit about the Sun. A combination of these effects means that the sub-solar latitude of Triton varies between 55° N and 55° S, with a period somewhat longer than Neptune's 165-year orbit because Triton's orbital plane precesses as well, meaning that Triton experiences exceptionally long seasons. When Voyager 2 flew by, the southern hemisphere was part-way through its spring season, with the fringe of its polar cap receding as the nitrogen ice sublimed and added to the atmosphere. Presumably the northern polar cap, in darkness and unseen, was growing as northern autumn progressed towards winter and nitrogen froze out of the atmosphere and on to the cold ground.

Where visible beyond the polar cap and through the thinner nitrogen frost, the surface is a mixture of water ice and carbon dioxide ice, with traces of methane and carbon monoxide. Ammonia is suspected, particularly in view of the cryovolcanic nature of the landscape, but has not been detected.

Triton's terrain is crossed by a number of furrowed ridges, known as sulci (Latin for 'grooves'), which are reminiscent of the largest features on Europa such as the two that intersect near the top of Figure 15. These may sit above fractures that were opened and closed (or possibly slid sideways) under the influence of tidal forces before Triton's orbit became virtually circular. In the lower part of Figure 22, lengths of some sulci are buried by smooth plains that look as if they were emplaced as cryovolcanic lava flows. Smooth, presumed cryovolcanic plains occur also in two irregularly shaped depressions near the middle right of Figure 22.

The terrain in the upper part of the figure is older. Here the sulci are uninterrupted, except where a younger sulcus cuts an older one, and they cross terrain pockmarked by curious 30–40 km dimples. No one knows what caused these. This is called 'cantaloupe terrain', because of its resemblance to a melon skin, but that doesn't help much! The dimples could overlie places where individual pods of 'warm' ice have risen through the crust, but regardless of their exact nature they are surely another instance of cryovolcanic activity.

22. **Voyager 2 mosaic using the best images of Triton. At the time of imaging, the fringe of the south polar cap was at 5–20° S. This view extends from about 60° S to 40° N.**

Triton's cryovolcanism and the stressing of the crust demonstrated by the sulci were presumably powered tidally, during the perhaps billion years that it took for Triton's orbit to become circular after its capture by Neptune. We don't know how long ago that was, but the impact craters scattered across the surface suggest that this activity has now ceased, or at least waned.

However, there are still eruptions of a kind on Triton. These occur through the polar cap, and were seen in progress (though imaged rather poorly) by Voyager 2. What seems to happen here is that sunlight passes through the largely translucent nitrogen ice and warms the darker substrate below. This warmth is conducted up into the nitrogen ice, until its base begins to sublime. A blister of nitrogen gas swells up beneath the icecap until it bursts, allowing the gas to jet skywards, taking with it dusty grains from the dark substrate. The plume becomes neutrally buoyant at a height of several kilometres. It is then blown downwind, and the dark particles begin to fall out, leaving low-albedo streaks that can be seen on the polar cap. If this explanation is correct, these are solar-powered geysers rather than true cryovolcanic eruptions. Nevertheless, they qualify Triton for membership of the exclusive club of moons with proven present-day activity that otherwise includes only Io, Titan, and Enceladus.

Triton is a wonderful world, but it won't be around for ever. Its orbit is already closer to Neptune than the Moon is to the Earth, and tidal processes to do with its inclined orbit passing over the planet's equatorial bulge are gradually dragging it closer. Calculations suggest that Triton will pass within Neptune's Roche limit in about three billion years, which could give birth to a ring system far more spectacular than Saturn's.

Future missions

It should be clear that there is much more to find out about even the best-known moons of the giant planets, and that more missions will be needed to deliver this knowledge. The earliest mission currently approved by any space agency is the ESA's Jupiter Icy moons Explorer (JUICE), planned for launch in 2022 and arrival for a thirty-month tour at Jupiter in 2030. This will begin with a series of close fly-bys of Europa and Callisto, using assists from their gravity until eventually the probe can be captured into orbit about Ganymede, into which it will eventually crash. Also for launch sometime in the 2020s is NASA's unimaginatively named Europa Mission. Like JUICE, this will orbit Jupiter but make dozens of flybys of Europa to study its subsurface ocean and to look for eruption plumes. A subsequent NASA or ESA mission, not yet funded, might usefully drop a series of penetrators or tiny 'chipsats' into the ice beside young cracks in Europa, to probe the ice and search for life.

Titan is an appealing target as well—and proposals for splashdown into one of Titan's largest methane seas are under consideration at NASA. I hope that such a mission would also get a chance to analyse Enceladus plume material for biological signatures. New missions to Triton and the moons of Uranus, of whose surfaces we have tantalizingly seen less than half, are lower down most priority lists.

Chapter 6
The moons of Mars: captured asteroids

Mars is the planet most recently found to have moons. In his fictional satire of 1726, *Gulliver's Travels*, Jonathan Swift described the astronomers of the flying island of Laputa as having discovered two small moons of Mars: 'They have likewise discovered two lesser stars, or satellites, which revolve about Mars; whereof the innermost...revolves in the space of ten hours, and the latter [outermost] in twenty-one and a half' (Part III, Chapter 3).

In reality it was a century and a half later, in 1877, that the moons of Mars were actually discovered, two in number as Swift has guessed, by the American Asaph Hall after a protracted search using what was then the world's largest refracting telescope (66 cm). He found the smaller, outer one (Deimos) first, and the larger, inner one (Phobos) six days later. Their actual orbital periods (see Appendix) are 7.7 hours and 30.2 hours respectively.

Swift's guess that Mars has two moons was probably an attempt to fit Mars into a pattern, bearing in mind that the Earth has one moon and (at the time) Jupiter had four known moons and Saturn five. Kepler made a similar numerical speculation in 1610, inspired by Jupiter's four moons before any of Saturn's had been discovered. The short orbital periods that Swift gave to his invented moons indicates his awareness that if any moons of Mars had longer periods, they would be far enough from the planet to have been discovered already by astronomers in the real world.

Both of them are small rocky bodies in synchronous rotation. Phobos is about the size of Saturn's Trojan moon Calypso (Figure 10) and Adrastea, the smallest inner moonlet of a giant planet included in Figure 12. Phobos measures 27 × 22 × 18 km, and Deimos 15 × 12 × 11 km. They are far too small for their own gravity to pull their shapes into hydrostatic equilibrium. Despite the fact that their orbits are prograde, only slightly eccentric, and inclined at only about 1° to Mars' orbital plane, they are almost certainly captured asteroids. This makes them more analogous to the irregular moons of the giant planets than to their inner moonlets. They have little in common with the Moon.

Their densities, just less than twice that of water, are too low for them to be solid rock. There may be some ice inside, though spectroscopic studies show no signs of hydration at the surface. It is more likely that their low densities, a property they have in common with most of those asteroids whose densities have been determined, are because below the surface regolith their interiors consist of chunks (unknown in size) of loosely packed rubble. This is a similar explanation to that offered for Saturn's Hyperion, an icy moon whose density is too low to be solid ice. Spectroscopically, both Phobos and Deimos resemble asteroids that are believed to equate to a class of meteorite known as carbonaceous chondrites.

Phobos

Phobos (Figure 23) has been imaged in better detail than Deimos, because its orbital height of only 5,645 km above the surface (as opposed to Deimos' 20,000 km altitude) brings it closer to spacecraft orbiting Mars, of which there have been several at typical altitudes of about 300 km. The Russian space agency has attempted three missions to Phobos. Of the duplicates Phobos 1 and 2, launched in 1988, only Phobos 2 made it to Mars, but contact was lost in January 1989 as it closed in on Phobos itself to attempt a landing. More recently, Phobos-Grunt, which was intended to bring back a sample from the surface, malfunctioned and failed to leave Earth orbit after launch in 2011.

Phobos has numerous craters. The younger ones are sharp but older examples are very muted, as if buried by regolith. Using the first close-up images of Phobos from the Mariner 9 Mars orbiter, seven craters were named in 1973: six after astronomers who had worked on Mars, including Hall himself. The seventh and largest (9 km in diameter) was named Stickney, after Angeline Stickney, Hall's wife, to whose encouragement, when he had been on the point of giving up his search, he attributed his eventual success in finding Mars' moons. Stickney occupies almost the whole of the lower-left end of Phobos in Figure 23, and is not far from the middle of Phobos' leading hemisphere. Although large in comparison to Phobos, Stickney is scarcely bigger than the lunar crater Copernicus B in Figure 4. It is far too small to have developed a central peak, and has a simple bowl shape. There is a prominent 2 km crater within it named Limtoc, one of eight names assigned in 2006 that were taken from *Gulliver's Travels* (Limtoc being the name of a Lilliputian general).

Individual craters on Phobos are surely the results of impacts, but there is a great deal of controversy about the 100–200-metre-wide grooves that can be seen crossing the surface. In detail, some are simple grooves whereas others are chains of overlapping pits. Many grooves appear to emanate (or radiate) from Stickney, though a few cross Stickney's rim and so must be younger than this crater. Families of grooves of different orientations can be recognized (especially beyond the area shown in Figure 23), with consistent age relationships exhibited by the overprinting of older-family grooves by younger-family grooves.

23. The Mars-facing side of Phobos, recorded by the HiRISE camera of NASA's Mars Reconnaissance Orbiter with a resolution of about six metres per pixel from a range of 5,800 km.

Early hypotheses to explain the grooves tried to relate them to Stickney. Could they be chains of overlapping secondary craters made by ejecta flung out from Stickney? This doesn't work, because any ejecta from Stickney travelling fast enough to make such craters would escape from Phobos' low gravity and so could not come back to strike the surface, and in any case some grooves are too young. Could they be fractures caused by the

Stickney impact? This falls down because of the various ages and orientations of groove families.

The near-radial disposition of most grooves with respect to Stickney is now mostly dismissed as a red herring, with Stickney being a coincidental impact near Phobos' leading point. If that is correct, the relationship that actually needs explaining is the one between the groove pattern and Phobos' orbital travel. There are two competing hypotheses based on this. One has the grooves as the surface manifestations of fractures that were imposed on Phobos' interior by tidal forces, or by aerodynamic drag from an extended ancient Martian atmosphere during capture of Phobos by Mars. Sets of fractures of the observed orientations could be formed in this way, and later widened to form the grooves. This is plausible, but the exactly planar fractures implied by the straightness of the grooves is hard to reconcile with the likely rubbly and porous interior required by Phobos' low density.

The other motion-based hypothesis notes that Phobos' low orbit must inevitably from time to time carry it through a hail of ejecta flung out by large crater-forming impacts on Mars. Fragments of this ejecta could hit Phobos either while still travelling upwards or when falling back down, which would account for the grooves on the side of Phobos that never faces the planet. Each groove would represent Phobos' passage through a string of ejecta, leaving a trail of damage on its surface like that on the bodywork of a car that had been driven through the fire of a machine gun. Several strings of ejecta from the same impact on Mars would be encountered at once, which is why the grooves come in families. There are objections to this explanation too, notably the remarkable lack of dispersion of the ejecta travelling from Mars required to make such straight grooves (i.e. each component of a string of ejecta would have had to travel in *precisely* the same direction, with no sideways deviation), so perhaps it is best to regard the grooves of Phobos as a mystery that has yet to be adequately explained.

Deimos

Deimos (Figure 24) has no grooves, probably not so much because it lacks any crater of the size of Stickney, but because it is further from Mars. This makes it less vulnerable to tidal or capture-related stresses and also less likely to encounter intense hails of ejecta from the Martian surface.

Deimos has a cratered surface similar to the smoother parts of Phobos. Only two of its craters have been named (both in 1973, based on Mariner 9 images): Swift (1 km in diameter, identified by the converging arrows in Figure 24) and Voltaire (1.9 km in diameter), which is a more subtle, less prominent feature immediately below Swift in Figure 24. The crater Voltaire is named after the French Enlightenment writer who, a quarter of a century after Swift, set down his own opinion that Mars must have two moons.

Moons in the Martian sky

Phobos and Deimos are the only moons in our Solar System (apart from our own) that could one day be admired in the sky by humans standing on the surface of the relevant planet, because Mars is the only planet with moons which has a surface that could be landed on. Being so much closer to Mars than the Moon is to the Earth, and in low-inclination orbits, they pass into Mars' shadow at night more often than lunar eclipses occur on Earth. Similarly, by day they often cross the disc of the Sun, but despite being close to Mars their size is too small to hide the whole of the Sun's disc, so there are no total solar eclipses on Mars.

When overhead, Phobos would look almost half the size of the Moon in the Earth's sky, but when close to the horizon it would be nearly twice as far away from the observer and so would look proportionately smaller. This effect is not noticeable for the Moon, whose orbit is much bigger than the Earth's radius.

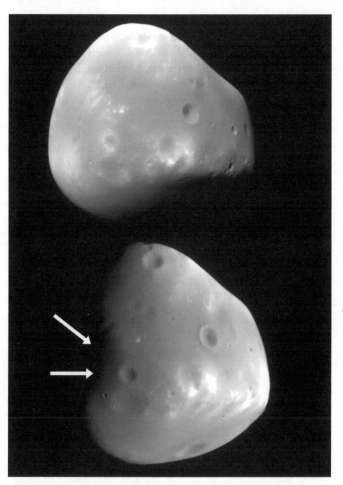

24. Two views of the same (Mars-facing) face of Deimos under different illumination, recorded by the HiRISE camera of NASA's Mars Reconnaissance Orbiter with a resolution of about twenty metres per pixel. The crater Swift is indicated by the converging arrows.

Phobos travels across the Martian sky faster than Deimos, and seen from low latitudes on Mars it regularly passes in front of it, presenting the remarkable sight of one moon hiding another. All of these phenomena have actually been imaged by cameras carried by rovers on the Martian surface, and I have included a link to a video of Phobos passing in front of Deimos in the Further reading.

Phobos and Deimos won't continue their orbital dance indefinitely. Phobos is so close to Mars that tidal forces are making its orbit decay much faster even than Triton's. Its nearest point to Mars is getting lower by about two centimetres each year, and at that rate it can survive for about only fifty million years more before being ripped apart or crashing to the surface.

Chapter 7
Moons of small bodies

The small bodies of our Solar System can be summed up as: asteroids (rocky or carbonaceous objects that are concentrated in, but not confined to, the space between the orbits of Mars and Jupiter); trans-Neptunian objects (icy bodies beyond Neptune's orbit, including Pluto); and comets (small icy bodies with strongly elliptical orbits that can come close to the Sun). These three types are useful distinctions, but not all objects can be neatly pigeonholed, as you will see shortly when a fourth type, centaurs, comes into the story.

Among all these, only comets are devoid of known moons. Comets have been known since antiquity only because they can develop spectacular tails of gas and dust when they pass close to the Sun. The solid part, or 'nucleus', of the largest comets is much smaller than the largest examples of any other kind of small body. Most comet nuclei are less than 10 km across, and there is no known comet nucleus as large as 100 km.

Three comet nuclei have been shown to have shapes consistent with being 'contact binaries', that is two main lumps loosely held together beneath a regolith layer, but no comet has been found to have a freely orbiting moon. Indeed, if a comet did have a moon it would be unlikely to survive for long, because vented gas would make their mutual orbits chaotic.

Asteroids with moons

Well-characterized asteroids are formally designated by a name preceded by a number corresponding roughly to the sequence of discovery. The first asteroid to be found was 1 Ceres (in 1801), which is also the largest at 950 km in diameter. Those that have not been tracked long enough to be sure of their orbit are unnamed, and have only a provisional designation consisting of the year of discovery followed by a two-letter (or two-letter-plus-number) code. Asteroids as small as about ten metres in size can be tracked by radar when they pass close to the Earth, but may never be seen again.

In 2015, 184 asteroids were known to have moons. Of these, 104 are in the main asteroid belt between the orbits of Mars and Jupiter, 21 pass inside Mars' orbit, and 55 cross or come close to Earth's orbit. Most of them have only one moon, but nine have two moons.

This remarkable tally is thanks mainly to advances in optical telescope imaging over the past twenty years. However, the first asteroid moon was discovered by accident in February 1994 when images from Galileo's fly-by of the asteroid 243 Ida six months previously were downloaded. These revealed a 1.5 km diameter moon that was later named Dactyl, whose cratered surface demonstrates that it is fairly old. It has a prograde orbit about Ida and apparently similar composition (Figure 25).

Dactyl remains the only moon of an asteroid to have been seen at close quarters, and to explain it we need to consider Ida's history. Ida is a member of the Koronis family of main-belt asteroids, which share similar orbits and are thought to result from a collision between two larger bodies about two billion years ago. There are twenty Koronis family asteroids like Ida that exceed 20 km in diameter and nearly 300 known smaller ones. Dactyl

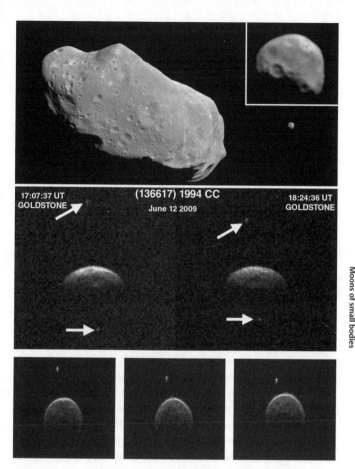

25. Asteroid Ida (56 km long) and its 1.5 km moon Dactyl in a single Galileo image (top), the inset is the highest-resolution view of Dactyl, enlarged, seen from a range of 3,900 km; views of 700-metre asteroid 1994 CC and its two tiny moons (arrowed), imaged by radar at a range of about 3,000,000 km in 2009 (middle); three successive radar views of asteroid 2004 BL_{86} and its seventy-metre moon, at a range of only 1.2 km from Earth on 26 January 2015 (bottom).

may be a fragment from the same parent body that broke away from Ida during this collision, but with too little force to become independent, or it could be a chunk knocked off Ida more recently.

All the other known moons of main-belt asteroids have been discovered telescopically, beginning in 1998 with Petit-Prince, a 13 km diameter moon taking 4.8 days to orbit the 213 km diameter asteroid 45 Eugenia at a distance of 1,184 km, close to its equatorial plane. Discovery of a 6 km diameter, closer-in, second moon of 45 Eugenia was announced in 2007. This has yet to receive an official name, but I understand that the name 'Princesse' is before the IAU for approval.

Before the second moon of 45 Eugenia had been discovered, the 286 km asteroid 87 Sylvia had already been found to have two moons, about 18 km and 7 km across and orbiting very close to Sylvia's equatorial plane at distances of 707 km and 1,357 km. These have been given the highly appropriate names Romulus and Remus (after the wolf-suckled twins in the story of Rome's founding) whose mythological mother was Rhea Sylvia after whom 87 Sylvia was named.

The examples so far have all been of a sizeable asteroid with either one or two much smaller moons, but 90 Antiope is something different. It was discovered in 1866, and seemed to be an unremarkable largish asteroid until the year 2000 when it began to be revealed as a binary object. We now know that it is two bodies, each about 86 km in size and 170 km apart, in synchronous rotation and orbiting their common centre of mass (or 'barycentre') with a period of 16.5 hours. Being indistinguishable in size or mass, each can be regarded as a moon of the other.

Much smaller asteroids can be seen only when they come close to the Earth, and it is even possible to study them by radar when less than about ten million kilometres away. Of the 'near-Earth asteroids', 16 per cent have been shown to be double or triple systems. For example, 69230 Hermes consists of two nearly identical objects about 400 metres across and only 1,200 metres apart, making them a miniature version of 90 Antiope. Other examples look like scaled-down versions of Ida/Dactyl or Sylvia/Romulus/Remus. Among these, the 2.8 km diameter 1998 QE_2 has been shown to rotate in about five hours, and its 600-metre-wide moon is in synchronous rotation in a thirty-two-hour orbit at a range of about 6 km. The asteroid 1994 CC (Figure 25) has two even tinier moons. The inner is about 110 metres across and seems to be in synchronous rotation in a 1.7 km radius orbit taking about thirty hours, whereas the outer (about eighty metres in diameter) seems to rotate faster than its nine-day orbital period. This was the smallest known moon of any object until the 355-metre asteroid 2004 BL_{86} passed Earth at a range of 1.2 million km in January 2015 and radar showed a moon only seventy metres across (Figure 25).

Many asteroid moons probably originated as debris knocked off the main body by an impact, but in the case of the smaller asteroids their moons could have begun as loosely held components flung off the main body by a centrifugal process as rotation speed increased. This surprising spin-up can happen because of something known as the YORP (Yarkovsky–O'Keefe–Radzievskii–Paddack) effect, whereby the angular momentum of a small, irregular rotating body is unbalanced because of absorption of sunlight on the dayside only, but emission of thermal radiation from the warmed surface that extends into the nightside.

Centaurs with moons

Centaurs are asteroid-like bodies of dominantly icy rather than rocky composition, whose orbits lie beyond that of Jupiter but inside Neptune's. About 200 are known, of which the largest example is smaller than the largest Kuiper-belt objects that are found beyond Neptune.

Moons have been imaged telescopically for two objects that are sometimes classified as centaurs: these are 42355 Typhon (162 km) and its moon Echidna (90 km), and 65489 Ceto (200 km) and its moon Phorcys (170 km). However, although these come inside the orbit of Uranus when at their closest to the Sun, their orbits are strongly elliptical and take them way beyond Neptune when at their most distant, so they are perhaps more appropriately regarded as inwardly scattered trans-Neptunian objects.

For an example of an uncontested centaur that may have moons, we need to turn to 10199 Chariklo. This is the largest known centaur, and has a diameter of about 250 km. In June 2013 it was predicted to pass exactly in front of a star (a rare event for an object so small) as seen from various South American observatories. Exact measurement of the interlude (typically a few seconds long) when the star is hidden, known as an occultation, provides the best way we have to work out the size of such a small distant object, so every telescope in the path of the occultation (which had the same width as Chariklo itself) was looking. To everyone's surprise, the star was dimmed briefly twice before and again twice after the main event. The only feasible explanation for this is that Chariklo has two rings: a 7 km wide ring at 391 km radius and a less dense 3 km wide ring at 405 km.

Chariklo is by far the smallest body found to have rings. These rings are very narrow, so unless they formed less than a few thousand years ago (which would be a remarkable coincidence) they are probably confined by the presence of kilometre-size, unseen shepherd moons.

Trans-Neptunian objects and their moons

When Pluto's largest moon was discovered, in 1978, Pluto was still regarded as a planet, as it had been ever since its discovery in 1930. However, Pluto was always a misfit, because it passes temporarily inside the orbit of Neptune, which is 10,000 times more massive. The two are in 2:3 orbital resonance about the Sun (Pluto orbits twice in the time it takes Neptune to complete three orbits), but they can never collide because whenever Pluto is at its closest to the Sun, Neptune is either about 50° ahead or about 50° behind. Moreover, Pluto's orbit is inclined in such a way that it is well above the orbit of Neptune when closest to the Sun, and in fact the distance between Pluto and Neptune always exceeds seventeen times the Earth–Sun distance.

In 1992, a second Pluto-like trans-Neptunian object was discovered, eventually named 136199 Eris, and now believed to be about 27 per cent more massive than Pluto although fractionally smaller in size. By 2015 the tally of trans-Neptunian objects exceeded 1,500, of which about eighty (including Pluto and Eris) are known to have moons. Many of these are in Pluto-like orbits in a band thirty to fifty times further from the Sun than the Earth's orbit. This region is referred to as the Kuiper belt, after Gerard Kuiper (1905–73) who predicted something similar. Beyond lies the 'scattered disc' of similar objects in more eccentric, and often more steeply inclined, orbits that take them out beyond a hundred Earth–Sun distances.

Pluto is one of the largest, but is not exceptional and is not distinct from the continuum of trans-Neptunian objects other than having been the first to be discovered, by some sixty years. This new knowledge of the Solar System made it illogical to continue to regard Pluto as a planet, unless Eris and several others were also to be classified as planets. At its annual meeting in Prague in 2006 the IAU voted to adopt a definition of planet that excluded bodies that do not greatly out-mass objects sharing similar orbits and that cross the orbit of a much more massive object. This demoted Pluto while leaving the status of the other eight planets unchanged.

This decision was (and still is) unpopular in some quarters, but was in my view the right one, and much less messy than any alternative. Those objects orbiting the Sun that do not qualify as planets but that are massive enough for their gravity to pull their shapes into hydrostatic equilibrium are classified as dwarf planets. This class includes one rocky asteroid (1 Ceres), and at least five icy trans-Neptunian objects of which Pluto is one. This definition sounds logical, but in practice it cannot always be applied with confidence, because only 1 Ceres and Pluto have been seen clearly enough to demonstrate their shapes. The others are merely assumed to have shapes in hydrostatic equilibrium on the basis of their estimated size and mass.

When Pluto's first moon was discovered, by the American James Christy (1938–) in 1978, using a 1.55-metre telescope, it was detectable merely as a bulge that was sometimes present on the side of blurred images of Pluto, though newer telescopes and modern techniques soon enabled the two bodies to be resolved. Christy proposed the name Charon for this moon, and this was accepted by the IAU. Christy, it is said, would have liked to name his discovery after his wife, Charlene, known as 'Char'. He knew that the IAU would not accept this, but realized that Charon would seem appropriate because in Greek mythology Charon was the ferryman who conveyed the souls of the dead across the river Styx to Hades, of which Pluto (the Roman name) was the ruler. The orthodox way to pronounce the ferryman's name is 'Kairon' but Christy, and most Americans, say 'Shairon' or 'Sharon', preserving the 'sh' sound of Charlene.

Pluto is now known to have five moons (see Appendix), of which Charon is by far the largest. Its mass is one-eighth that of Pluto, so the two bodies are much more similar than the Earth and Moon, where the mass ratio is 1:80. Because of this, their barycentre is not inside Pluto but occurs about one Pluto radius above its surface, in the direction of Charon. The two bodies mutually orbit this point, and are tidally locked into synchronous rotation so that there is one hemisphere of Pluto from which Charon can never be seen, as well as one hemisphere of Charon from which Pluto can never be seen. There are also, of course, longitudes on Pluto experiencing eternal moonrise or moonset.

Pluto's axis is inclined at 119.6° to its orbit (so, like Uranus', its rotation is retrograde). All its moons' orbits lie within a fraction of a degree of its equatorial plane, and are prograde (travelling in the same direction as Pluto's spin) and nearly circular. The orbital periods from inner (Charon) to outer (Hydra) are nearly (but not quite) in 1:3:4:5:6 resonance.

The names of the smaller moons follow the underworld theme begun by Pluto and Charon: Nix, the goddess of darkness and the night, was the mother of Charon (the ferryman, not Mrs Christy); Hydra was a nine-headed serpent, and some say that this is a reminder of Pluto's temporary reign as a ninth planet; Styx is named after the goddess of the river across which Charon the ferryman plied his trade; and Kerberos was the three-headed dog that guarded the entrance to the underworld. Similar underworld names were in reserve for any other moons discovered by the New Horizons probe that flew past in July 2015, but it did not find any. However, during an all-too-brief encounter at a speed of 14 km/s, passing about 10,000 km from Pluto and 27,000 km from Charon, it revealed large regions of their globes in glorious detail, with a smallest pixel size less than 200 m.

Charon may have formed as the result of a 'giant impact' event similar to that proposed for the Moon's origin, in which case the four smaller moons could be debris from the same event that has migrated outwards. If Nyx and Hydra turn out to be in synchronous rotation this will suggest that outward migration has occurred, because calculations show that tidal forces would be inadequate to achieve this at their current distances. If they are not in synchronous rotation, this will suggest capture (with all its attendant difficulties) as a more likely origin.

Pluto is covered by ices of nitrogen, methane, and carbon monoxide, overlying stronger and less-volatile 'bedrock' made of water-ice, but is dense enough that it must have rock in its deep interior. It has an atmosphere derived from its surface ices, dominated by nitrogen. In all these respects it appears to resemble Neptune's large moon Triton, which you may recall is likely to have been captured from the Kuiper belt. New Horizons close-up images found Pluto's surface to be even less-cratered than Triton's, showing that it has been resurfaced even more recently. Some high resolution frames seem to contain no craters at all, suggesting a local surface age of less than 100 million years. Triton-like sulci and cantaloupe terrain are absent. Instead, there are tracts where water-ice sticks up through the other ices to form jagged 3 km high mountains, and plains of carbon monoxide ice whose surface looks like a larger scale version of ground patterned by freeze-thaw process in the Earth's arctic and parts of Mars.

In contrast, Charon's surface is dominated by water ice, but hydrated ammonia was detected telescopically even before New Horizons, prompting speculation of recently active cryovolcanism. The amount of tidal heating in the Pluto–Charon system is not well understood, but the varying effect of the Sun's pull on the steeply inclined Pluto–Charon orbit, to which Charon is more vulnerable than Pluto, might be enough to generate the heat required to partially melt ice mixtures in their interiors. First thoughts from the New Horizons team were that the amount of recent heating required to account for Pluto's terrain could not be explained tidally. Although they could cite no viable alternative mechanism, they speculated that this cast doubt upon the tidal heating explanation for young surfaces on icy bodies elsewhere too. A simpler, but equally contentious explanation is that Charon was captured, or formed by a giant impact, in the relatively recent past rather than billions of years ago.

Charon too has a youngish surface (Figure 26), though there are fewer signs of recent activity than on Pluto. It is notable for faulted valleys in the form of straight-sided troughs a few kilometres deep similar to those seen on Uranus's moons Ariel (Figure 21) and Titania, and a dark north polar cap that the New Horizons team named 'Mordor Regio'.

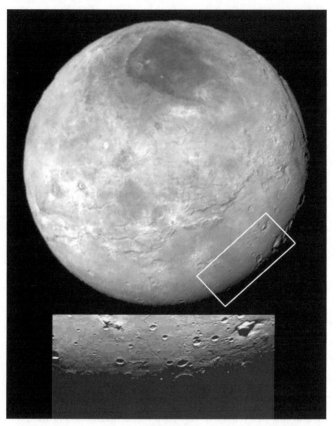

26. A global view of Charon by New Horizons. Note the north–south fracture system close to the eastern limb (right hand edge) and a second major fracture system running close to the equator. The dark polar region is Mordor Regio. Impact craters near the terminator are revealed by their interior shadows. Elsewhere they show up because of surrounding bright ejecta deposits, or sometimes because they have excavated darker material. The box locates the area shown in a more detailed image below (recorded several hours later so the terminator has migrated). This shows a perplexing mountain peak surrounded by a moat, and at the limit of resolution several straight fractures reminiscent of those west of Jansen E in Figure 5. There they are extensional fractures in mare basalt attributed to shallow intrusion on a vertical curtain of magma (a dyke), on Charon they could be caused by similar intrusion of a cryomagma made of a water-ammonia mixture.

As well as mapping the landforms and compositions of Pluto and Charon, New Horizons looked back at the Sun and the Earth as they were occulted in turn by each body to learn about Pluto's atmosphere and to try to determine whether Charon has any atmosphere at all. The smaller moons were imaged well enough to reveal their shapes, which are irregular as expected.

In 2018 or 2019 New Horizons will make a fly-by of at least one other Kuiper-belt object and its moons (if any). Moons of trans-Neptunian objects in general come in many combinations, and it is likely that most owe their origins to collisions. The object 136199 Eris bears the name of the ancient Greek goddess of strife and discord, a reference to the hornet's nest that its discovery stirred up regarding Pluto's status. It has a single known moon discovered in 2005 that has been allocated the name Dysnomia, which was borne by Eris' mythological daughter. Eris has a radius of about 1,200 km, whereas Dysnomia's must be somewhere in the region of 100–300 km, depending on its albedo. Dysnomia's orbit proves that the mass of Eris is significantly (about 27 per cent) larger than Pluto's, although its size is slightly smaller. Eris and Dysnomia are challenging objects to observe, because they are currently three times further from the Sun than Pluto is.

Size estimates of such distant objects are very uncertain, because they are mostly too small and too far away to show resolvable, measurable discs in a telescope. If a stellar occultation is observed and its duration measured, you can use the body's orbital speed to deduce the size of the part of the body that passed in front of the star, though that is likely to be a chord rather than a full diameter. These are rare events, which is why observatories were geared up to observe the stellar occultation by 10199 Chariklo that fortuitously revealed its rings. In the absence of any occultation data, if you assume an albedo you can use an object's brightness to deduce size. The uncertainties in this method meant that we were for a while unsure whether Eris was larger or smaller than Pluto, even though we were already sure that Eris was more massive.

Pluto and Charon's respective sizes were initially deduced in a series of mutual occultations that occurred between 1985 and 1990 when the plane of their orbits crossed the line of sight from Earth. A similar series of events between July 2014 and October 2018 should reveal the true sizes of the approximately 80 km Kuiper-belt object 385446 Manwë and its approximately 50 km moon Thorondor (whose names derive from the works of J. R. R. Tolkien). This is of more than passing interest because the current best estimate of Manwë's size means that its density is less than that of water, in which case it must be an icy rubble pile. Manwë's orbit round the Sun is in 4:7 resonance with Neptune's, as opposed to Pluto's 2:3 resonance.

One of the largest Kuiper-belt objects is 136108 Haumea , with a mean radius estimated at about 650 km. Its brightness goes up and down in a little less than two hours, revealing it as an elongated body with a 3.9-hour rotation period. The remarkably fast spin is believed to be responsible for distorting it into an elongated ellipsoid rather than merely bulging at the equator, like a slower-spinning body would. Its pole-to-pole diameter is probably just under 1,000 km, whereas the maximum and minimum diameters across its equator are about 1,960 and 1,520 km. It is dense enough to have a rocky interior, but its surface spectrum is that of crystalline water ice.

Haumea has two known moons, Hi'iaka (340 km diameter, orbiting at 49,900 km in 49.5 days) and Namaka (170 km diameter, orbiting at 25,700 km in 18.3 days). Like Haumea itself, these have the spectroscopic properties of crystalline water ice and a likely high albedo. Because of their shared properties, origin by capture is improbable and Haumea's moons are probably fragments that were flung off during an interlude of even more rapid spin or from a collision. Haumea is named after of a matron goddess of Hawai'i, whose many children (including Hi'iaka and Namaka) sprang from various parts of her body, as indeed the moons of the same names may have done.

The only other trans-Neptunian triple system apart from Haumea occurs in a Pluto-like orbit (2:3 resonance with Neptune) and is known as (47171) 1999 TC_{36}, not yet having been formally named. It consists of a central binary object, with each component about 260 km in diameter, so we are in the same dilemma as for Antiope as to which is the moon of which. However, in this case the binary pair is orbited by a 140 km moon. The average density of the two binary objects, estimated without the aid of mutual occultations to confirm their sizes, appears to be only about two-thirds that of ice. If this is correct, then each is probably an icy rubble pile.

Chapter 8
Moons in other planetary systems: exomoons

The first definite discovery of a planet around another star
(an 'exoplanet') was made in 1995. Twenty years on, we know of
more than 1,000 stars with exoplanets, of which nearly half have
more than one. It now seems likely that 20 per cent of Sun-like
stars have at least one giant exoplanet whereas at least 40 per cent
may have lower-mass exoplanets.

In our Solar System, moons are considerably more numerous
than planets, and it would be surprising if exomoons did not
outnumber exoplanets. However, they will be very challenging to
detect. Only a few exceptional exoplanets have been seen by direct
imaging, and any exomoons are presently well below the visibility
threshold. The vast majority of exoplanets are inferred either
by cyclical changes in their star's radial velocity as it orbits the
barycentre between itself and its exoplanet, or, in cases where the
exoplanet's orbital plane lies in our line of sight, by observing
the tiny dip in the starlight as the exoplanet transits across the
star's disc, and so blocks a small fraction of its light.

Of these two techniques, careful use of the transit method offers
the greatest hope of revealing an exomoon. If the exomoon passes
in front of the star before or after the exoplanet, this will cause a
tiny dip in the star's brightness just before or after the larger dip
attributable to the planet itself. Furthermore, even if the dip in

starlight caused by the exomoon's transit were not detectable, analysis of a series of repeated exoplanet transits might reveal fluctuations in their precise timing caused by the exoplanet's displacement to either side of the exoplanet–exomoon barycentre. An Io-like exomoon could be signalled by the presence of a plasma torus, whose passage across the face of the star could both dim its light and reveal itself through characteristic spectral lines. There might also be recognizable radio emissions that could be traced to such a plasma torus. Perhaps the strongest evidence for exomoons concerns a planet of an orange dwarf star known as J1407, revealed to be a 'super-Saturn' with a vast and spectacular ring system by multiple dips in the star's light as the rings transit across it. The gaps in the rings strongly suggest resonances with substantial moons, or the presence of less massive shepherd moons.

So far, none of these techniques has yielded definitive proof of exomoons. Colleagues working in the field of exoplanet research assure me that we may not be far off, though I expect that the first genuine detection of an exomoon may take some time to be confirmed beyond reasonable doubt.

Why do exomoons matter? Well, there are at least as many apparently habitable moons in our Solar System (Europa and Enceladus) as there are planets (Earth and Mars). If hydrothermal vents on ocean floors really are a good place for life to begin as suggested in Chapter 5, then icy exomoons with internal oceans throughout the galaxy could host life. As we have seen, the prospects for such life are reasonable even where the body's surface temperature is way below the freezing point of water.

Tidally heated icy exomoons could host life well beyond what has conventionally been regarded as a star's 'habitable zone', which requires liquid water at the surface. This may be mostly microbial life, and one might argue that it would be hard for any intelligent multicellular life to develop a technological civilization underwater, but it is feasible that there are many more inhabited moons than there are inhabited planets.

Internal oceans are not the only habitable environment that we can imagine for an exomoon. Exoplanet searches have revealed many giant exoplanets much closer to their star than Jupiter is to the Sun. Such a giant exoplanet could feasibly have an Earth-like exomoon, on which Earth-like life, and even intelligence, could arise. A famous example in recent science fiction is Pandora, the setting for the James Cameron movie *Avatar*, which orbits a giant exoplanet in its star's conventional habitable zone.

Maybe, if we ever do establish communication with aliens, they may be sceptical of our claim that we come from a world that goes directly round a star, rather than from one that orbits a giant planet.

Appendix: moons data

This table contains data for all the known moons of the planets and Pluto. Mean radius is quoted as a way to specify the size of a non-spherical moon by a single value.

Moons of the Earth and Mars

Name	Orbital radius/ 10^3 km	Orbital period/ days	Mean radius/ km	Mass/ 10^{20} kg	Density/ 10^3 kg m^{-3}
The Moon	384.4	29.53	1737	734.2	3.344
Phobos (Mars)	9.378	0.319	11.2	0.000106	1.90
Deimos (Mars)	23.46	1.26	6.1	0.000024	1.75

Jupiter's moons

Name	Orbital radius/ 10^3 km	Orbital period/ days	Mean radius/ km	Mass/ 10^{20} kg	Density/ 10^3 kg m^{-3}
Four inner	128–222	0.29–0.67	10–73	<0.075	<3.10
Io	421.6	1.769	1,822	893	3.50
Europa	670.9	3.551	1,561	480	3.01
Ganymede	1,070	14.97	2,631	1,482	1.94
Callisto	1,883	26.33	2,410	1,076	1.83
Fifty-nine outer	7,507–24,540	130–779	1–85	<0.095	

Saturn's moons

Name	Orbital radius/ 10^3 km	Orbital period/days	Mean radius/ km	Mass/ 10^{20} kg	Density/ 10^3 kg m^{-3}
Sixteen inner and co-orbital	134–377	0.56–2.74	16–93	<0.019	<1.300
Mimas	185.5	0.952	197	0.379	1.15
Enceladus	238.0	1.37	251	1.08	1.61
Tethys	294.7	1.89	528	6.18	0.985
Dione	277.4	2.74	561	11.0	1.48
Rhea	572.0	4.52	763	23.1	1.24
Titan	1,222	15.9	2,575	1,346	1.88
Hyperion	1,481	21.3	133	0.056	0.55
Iapetus	3,561	79.3	746	18.1	1.09
Thirty-eight outer	11,110–25,110	449–1,491	3–109	< 0.083	

Uranus' moons

Name	Orbital radius/ 10³ km	Orbital period/days	Mean radius/ km	Mass/ 10²⁰ kg	Density/ 10³ kg m⁻³
Thirteen inner	49.8–97.7	0.345–0.923	5–81		
Miranda	129.4	1.41	234	0.66	1.20
Ariel	191.0	2.52	578	13.5	1.67
Umbriel	266.3	4.14	528	585	1.40
Titania	435.9	8.71	561	789	1.71
Oberon	583.5	13.5	763	761	1.24
Nine outer	4,276–20,900	267–2,824	9–75		

Neptune's moons

Name	Orbital radius/ 10³ km	Orbital period/days	Mean radius/ km	Mass/ 10²⁰ kg	Density/ 10³ kg m⁻³
Six inner	48.2–105	0.294–0.950	18–102		
Proteus	117.6	1.12	208	0.5	
Triton	354.8	5.88	1,353	214	2.059
Nereid	5,513	360	170		
Five outer	15,730–48,390	1,880–9,374	20–30		

Pluto's moons (the sizes of some small moons are given as range estimates)

Name	Orbital radius/ 10^3 km	Orbital period/ days	Mean radius/ km	Mass/ 10^{20} kg	Density/ 10^3 kg m^{-3}
Charon	19.6	6.39	608	16.2	1.85
Styx	42	20.2	5–13		
Nix	48.7	24.9	21×18		
Kerberos	59.0	32.1	6–17		
Hydra	68.8	38.2	27×20		

Moons

Further reading

Books

If you want a book about planets at a similar level to this one, then I suggest:

D. A. Rothery, *Planets: A Very Short Introduction* (Oxford University Press, 2008).

There are many books about the Moon, and a few about some individual moons of other planets. I know of no recent book that attempts to describe moons in general in more detail than here. My own *Satellites of the Outer Planets: Worlds in Their Own Right*, 2nd edition (Oxford University Press, 2000) is now rather out of date but can still be found.

For the Moon in particular:

A. L. Chaikin, *A Man on the Moon: The Voyages of the Apollo Astronauts* (Penguin Books, 2007). At nearly 700 pages this is one of the most complete accounts of the Apollo programme.

A. Crotts, *The New Moon: Water, Exploration and Future Habitation* (Cambridge University Press, 2014). An excellent, recent account.

G. H. Heiken, D. T. Vaniman, and B. M. French (eds), *Lunar Sourcebook* (Cambridge University Press, 1991). A superb repository of lunar data and our understanding as it matured after Apollo. The complete book is viewable or downloadable free of charge at <http://www.lpi.usra.edu/publications/books/lunar_sourcebook/>.

S. Ross, *Moon* (Oxford University Press, 2009). A bulky 'coffee table' book, which has been favourably reviewed.

Books on other individual moons:

R. Greenberg, *Unmasking Europa* (Springer, 2008). A clear and authoritative account of Europa, with some scathing passages concerning the struggles to get the thin ice theory for Europa published in the face of establishment opposition.

R. Lorenz and J. Mitton, *Titan Unveiled: Saturn's Mysterious Moon Explored* (Princeton University Press, 2008). Published rather soon after Cassini's orbital tour of Saturn began, so check for a more recent specialist book on Titan.

M. Meltzer, *The Cassini–Huygens Visit to Saturn* (Springer, 2015). Mostly about mission planning and design, less than a quarter of the book is about the moons and rings.

J. Spencer and R. Lopes, *Io after Galileo: A New View of Jupiter's Volcanic Moon* (Springer, 2005). A thorough account published after the Galileo mission.

Online resources

There is a great deal of excellent and well-illustrated material about moons on the Internet, including galleries of images, short videos, and whole courses that you can study for free. The list below merely contains some of the highlights.

FutureLearn/Open University MOOC about moons (a free online course, three hours per week over eight weeks) <https://www.futurelearn.com/courses/moons>.

Virtually the same free course on an Open University site, to study at your own pace, but with less support <http://www.open.edu/openlearn/ science-maths-technology/moons/content-section-overview>.

A 'Moon trumps' card game, to play online and test your knowledge of moons against a computer <http://www.open.edu/openlearn/ moontrumps>.

Four Open University videos about moons <http://www.open.edu/ openlearn/science-maths-technology/science/across-the-sciences/ moons-the-solar-system>.

Moon myths debunked in a short animated video <http://www.open. edu/openlearn/moonmyths>.

Another debunking of supermoons and a demonstration of the 'Moon illusion' <http://www.universetoday.com/118990/ why-does-the-moon-look-so-big-tonight/>.

A video recording of a fifteen-minute lecture I gave about life in moons for Word Space Week in 2014 <https://www.youtube.com/watch?v=95RpUMHqw-0>.

A gallery of NASA images of all planets and their moons <http://photojournal.jpl.nasa.gov/>.

'Quasi-moon' horseshoe orbits and the orbit of Cruithne can be examined at <http://www.astro.uwo.ca/~wiegert/3753/3753.html>.

A radar video of the smallest known moon, orbiting asteroid 2004 BL_{36} <http://www.jpl.nasa.gov/news/news.php?feature=4459>.

High-resolution views of the Apollo landing sites, showing footprints and rover tracks from the Lunar Reconnaissance Orbiter <http://www.lroc.asu.edu/featured_sites/>. Use the 'Flip Book' option to see how shadows change during the lunar day. You can also see other landers and rovers, and some new craters.

See how the Moon must rotate once per orbit to keep the same face permanently towards the Earth at <http://www.open.edu/openlearn/RotatingMoon>.

Animations showing the Moon's phases, libration, and changing apparent size over the course of a year <http://svs.gsfc.nasa.gov/cgi-bin/details.cgi?aid=4236>.

You can find a movie of an eruption of Tvashtar on Io if you visit <http://pluto.jhuapl.edu/Multimedia/Videos/index.php> and search under 'Data Movies' and then 'New Horizons at Jupiter'.

A three minute video about eruptions on Io and Enceladus <https://www.youtube.com/watch?v=ZVybNKuhpSY>.

Images, and other products, from the Cassini mission exploring Saturn and its moons <http://saturn.jpl.nasa.gov/>.

Waltz Around Saturn, a movie made from successive Cassini images and brilliantly set to music <http://vimeo.com/70532693>.

A movie of Prometheus and Pandora shepherding Saturn's F ring <http://saturn.jpl.nasa.gov/video/videodetails/?videoID=95>.

Voyager Uranus flyby movie <https://www.youtube.com/watch?v=DrKQaDupdWQ>.

Voyager Triton flyby (based on processing in 2014) <http://www.lpi.usra.edu/icy_moons/neptune/triton/movie/index.shtml>.

Phobos passing across Deimos, as seen from the surface of Mars <http://photojournal.jpl.nasa.gov/catalog/PIA17089>.

The New Horizons website, showing fly-by images of Pluto's moons <http://pluto.jhuapl.edu/>.

An audacious proposal to scatter 'chipsats' onto Europa is described here, though without mentioning any planetary protection considerations <http://www.astrobio.net/news-exclusive/ swarm-tiny-spacecraft-explore-europas-surface-rapid-response>.

Index

Moons

N

O

P

Q

R

SOCIAL MEDIA
Very Short Introduction

Join our community
www.oup.com/vsi

- Join us online at the official Very Short Introductions **Facebook** page.
- Access the thoughts and musings of our authors with our online **blog**.
- Sign up for our monthly **e-newsletter** to receive information on all new titles publishing that month.
- Browse the full range of Very Short Introductions online.
- Read **extracts** from the Introductions for free.
- Visit our library of **Reading Guides**. These guides, written by our expert authors will help you to question again, why you think what you think.
- If you are a teacher or lecturer you can order inspection copies quickly and simply via our website.

ONLINE CATALOGUE
A Very Short Introduction

Our online catalogue is designed to make it easy to find your ideal Very Short Introduction. View the entire collection by subject area, watch author videos, read sample chapters, and download reading guides.

http://global.oup.com/uk/academic/general/vsi_list/

PLANETS
A Very Short Introduction
David A. Rothery

This *Very Short Introduction* looks deep into space and describes the worlds that make up our Solar System: terrestrial planets, giant planets, dwarf planets and various other objects such as satellites (moons), asteroids and Trans-Neptunian objects. It considers how our knowledge has advanced over the centuries, and how it has expanded at a growing rate in recent years. David A. Rothery gives an overview of the origin, nature, and evolution of our Solar System, including the controversial issues of what qualifies as a planet, and what conditions are required for a planetary body to be habitable by life. He looks at rocky planets and the Moon, giant planets and their satellites, and how the surfaces have been sculpted by geology, weather, and impacts.

"The writing style is exceptionally clear and pricise"

Astronomy Now

www.oup.com/vsi

STARS
A Very Short Introduction
Andrew King

Every atom of our bodies has been part of a star. Our very own star, the Sun, is crucial to the development and sustainability of life on Earth. This *Very Short Introduction* presents a modern, authoritative examination of how stars live, producing all the chemical elements beyond helium, and how they die, sometimes spectacularly, to end as remnants such as black holes. Andrew King presents a fascinating exploration of the science of stars, from the mechanisms that allow stars to form and the processes that allow them to shine, as well as the results of their inevitable death.

"Part of the extensive Very Short Introduction series, this volume by Andrew King provides an engaging overview of the science of stars. This pocket-sized book is an enjoyable read."

Dawn E. Leslie, Contemporary Physics

BLACK HOLES
A Very Short Introduction
Katherine Blundell

In this *Very Short Introduction*, Katherine Blundell addresses a variety of questions, including what a black hole actually is, how they are characterized and discovered, and what would happen if you came too close to one. She explains how black holes form and grow - by stealing material that belongs to stars, as well as how many there may be in the Universe. She also explores the large black holes found in the centres of galaxies, and how black holes give rise to quasars and other spectacular phenomena in the cosmos.

[there are no reviews]

www.oup.com/vsi

COSMOLOGY
A Very Short Introduction
Peter Coles

This book is a simple, non-technical introduction to cosmology, explaining what it is and what cosmologists do. Peter Coles discusses the history of the subject, the development of the Big Bang theory, and more speculative modern issues like quantum cosmology, superstrings, and dark matter.

"There is an embarrassment of books about the universe for the general reader, but few manage to pack so much, so elegantly, into such a compact space as this does. The book is generously illustrated."

The Guardian

"Coles takes you gently through everything from Blue Shift to parallel Universe in a thoroughly entertaining read"

TNT Magazine

www.oup.com/vsi

ASTROPHYSICS
A Very Short Introduction
James Binney

In this *Very Short Introduction*, the leading astrophysicist James Binney shows how the field of astrophysics has expanded rapidly in the past century, with vast quantities of data gathered by telescopes exploiting all parts of the electromagnetic spectrum, combined with the rapid advance of computing power, which has allowed increasingly effective mathematical modelling. He illustrates how the application of fundamental principles of physics - the consideration of energy and mass, and momentum - and the two pillars of relativity and quantum mechanics, has provided insights into phenomena ranging from rapidly spinning millisecond pulsars to the collision of giant spiral galaxies. This is a clear, rigorous introduction to astrophysics for those keen to cut their teeth on a conceptual treatment involving some mathematics.

[There are no reviews]

www.oup.com/vsi